Karsten Voigt

Interferometers in SOI-Technology for Use in Telecommunications

Karsten Voigt

Interferometers in SOI-Technology for Use in Telecommunications

Interferometers in SOI-Technology

Südwestdeutscher Verlag für Hochschulschriften

Impressum / Imprint

Bibliografische Information der Deutschen Nationalbibliothek: Die Deutsche Nationalbibliothek verzeichnet diese Publikation in der Deutschen Nationalbibliografie; detaillierte bibliografische Daten sind im Internet über http://dnb.d-nb.de abrufbar.

Alle in diesem Buch genannten Marken und Produktnamen unterliegen warenzeichen-, marken- oder patentrechtlichem Schutz bzw. sind Warenzeichen oder eingetragene Warenzeichen der jeweiligen Inhaber. Die Wiedergabe von Marken, Produktnamen, Gebrauchsnamen, Handelsnamen, Warenbezeichnungen u.s.w. in diesem Werk berechtigt auch ohne besondere Kennzeichnung nicht zu der Annahme, dass solche Namen im Sinne der Warenzeichen- und Markenschutzgesetzgebung als frei zu betrachten wären und daher von jedermann benutzt werden dürften.

Bibliographic information published by the Deutsche Nationalbibliothek: The Deutsche Nationalbibliothek lists this publication in the Deutsche Nationalbibliografie; detailed bibliographic data are available in the Internet at http://dnb.d-nb.de.

Any brand names and product names mentioned in this book are subject to trademark, brand or patent protection and are trademarks or registered trademarks of their respective holders. The use of brand names, product names, common names, trade names, product descriptions etc. even without a particular marking in this works is in no way to be construed to mean that such names may be regarded as unrestricted in respect of trademark and brand protection legislation and could thus be used by anyone.

Coverbild / Cover image: www.ingimage.com

Verlag / Publisher:
Südwestdeutscher Verlag für Hochschulschriften
ist ein Imprint der / is a trademark of
AV Akademikerverlag GmbH & Co. KG
Heinrich-Böcking-Str. 6-8, 66121 Saarbrücken, Deutschland / Germany
Email: info@svh-verlag.de

Herstellung: siehe letzte Seite /
Printed at: see last page
ISBN: 978-3-8381-3410-9

Zugl. / Approved by: Berlin, TU, Diss., 2012

Copyright © 2012 AV Akademikerverlag GmbH & Co. KG
Alle Rechte vorbehalten. / All rights reserved. Saarbrücken 2012

Contents

1	**Introduction**	**3**
2	**Mach-Zehnder delay interferometer for phase demodulation**	**7**
2.1	Modulation formats	7
2.2	General phase demodulator scheme	8
2.3	Ideal and Non-ideal MZ-DI – a description with matrix theory	10
2.4	Demodulator requirements	18
3	**4μm SOI rib waveguides**	**21**
3.1	SOI material system & single-mode condition	21
3.2	Optical Losses in SOI rib waveguides	23
3.3	Coupling single-mode fiber (SMF) - SOI rib waveguide	26
3.4	Fabrication of SOI rib waveguides	27
3.5	Measurement setup	30
3.6	Characterization of straight and bent waveguides	32
3.7	Polarization dependence of SOI rib waveguides	33
3.7.1	Geometrical birefringence	34
3.7.2	Stress induced birefringence	37
4	**Multi-mode interference (MMI) coupler**	**43**
4.1	Design of MMI couplers	43
4.2	Two selected MMI couplers	55
4.3	Experimental results	62
4.4	Fabrication tolerances of MMI couplers	65
5	**Experimental results of Mach-Zehnder delay interferometers**	**70**
5.1	Layout	70
5.2	2×2 MZ-DI characterization	72
5.3	2×4 MZ-DI characterization	77
5.4	Thermal tuning of MZ-DI	81
5.5	Impacts on MZ-DI birefringence	87

Applications on a 4µm SOI material platform..........89

6	**Hybrid integrated DPSK receiver** 89
6.1	Background 89
6.2	Layout and Flip-chip process..90
6.3	System Performance of a 40 Gb/s DPSK demodulator 95
6.4	System performance of a 40 Gb/s DPSK receiver module 98

7	**SOI material as platform for all-optical wavelength conversion** 100
7.1	Introduction 100
7.2	SOI-board for wavelength conversion................................100
7.3	AOWC with cascaded SOI MZ-DIs 103
7.4	AOWC with integrated SOA on SOI board........................105

8   Summary..111

9   List of acronyms 113

10   List of symbols 115

11   List of Publications 120

12   Bibliography...125

13   Acknowledgements 133

1 Introduction

Silicon-on-insulator (SOI) is an attractive material system to implement photonic applications as planar lightwave circuits (PLC). The last decade has seen an enormous development in SOI technology for passive devices and integration of active devices. Passive SOI waveguides show low intrinsic loss and low optical wavelength dependence. The large index contrast achievable in SOI technology attracts a lot of attention in applications with strong miniaturization. The shrinkage in waveguide dimensions (nanowires with dimensions of a few 100nm) supports the attempts to combine optical functionalities with microelectronics. This opens the door of SOI technology in microelectronic factories [1]. Furthermore, the integration of other materials could be demonstrated successfully [2; 3].

This PhD thesis is focused on the development of two applications using a 4µm-SOI technology [4; 5; 6]:

- 40 Gb/s differential phase-shift keying (DPSK) demodulator
- All-optical wavelength converter (AOWC) for λ - conversion of up to 160 Gb/s

Mach-Zehnder delay interferometer (MZ-DI) is the central components and building block for both applications. Its quality determines the performance of the applications. MZ-DI is a passive component. It consists of single-mode waveguides and multi-mode interference (MMI) couplers.

In terms of material dimensions, 4µm SOI belongs to the middle of the SOI spectrum. In comparison to the thicker SOI material (for example ~ 11µm used at TU-Berlin in the 90's [7; 8]) it enables the designing of smaller devices. The weak scattering loss in waveguides using this technology, avoids sophisticated fabrication requirements needed to realize photonic wires.

The good match between the modal fields of the single-mode (SM) fibers and SOI rib waveguides considerably minimizes the coupling loss in 4µm SOI waveguides. Furthermore, the birefringence of rib waveguides in 4µm SOI technology ($\sim 10^{-4}$) is considerably smaller than in nano-waveguides ($\sim 10^{-2}$).

A large part of this thesis will present theory, numerical analysis and experimental work that support optimization of 4µm SOI waveguide technology. Waveguide technology, design of high-performance MMI devices and the resulting high-performance MZ-DIs will be introduced.

MZ-DIs for phase-demodulation have to provide specific requirements in terms of polarization independence, which could only be realized using birefringence tuning techniques that will be outlined in this thesis. A considerable part of this thesis therefore deals with passive device technology.

For the above mentioned applications, the SOI material serves as an integration platform for the active InP-components. A balanced photo detector (BPD) has to be integrated for the DPSK-demodulator. The all-optical wavelength converter consists of cascaded SOI MZ-DIs together with an integrated semiconductor optical amplifier (SOA). This requires careful board design, which adapts active component integration as well as package requirements, i.e. predefined chip size, waveguide positions as well as bond pads.

SOI technology therefore serves as platform for higher levels of integration and shows a high potential for many optical applications. The presented results will underline that state-of-the-art performance can be achieved with SOI.

Organization of this thesis

Chapter 2 describes MZ-DIs for phase demodulation of advanced modulation formats, e.g. differential phase-shift keying (DPSK) or differential quadrature phase-shift keying (DQPSK) signals. A demodulator scheme for both formats is presented. The characteristics of Mach-Zehnder delay interferometers are described by matrices for ideal and non-ideal case. Furthermore, the chapter describes demodulator requirements with emphasis on polarization dependent frequency shift (PDFS).

Chapter 3 reviews the 4μm SOI rib waveguide characteristics, i.e. SM-condition and optical losses. The chapter continues with consideration of coupling loss between the standard optical fiber and the 4μm rib waveguide. Then, a short overview of rib waveguide fabrication at TU Berlin is provided. This includes the clean-room fabrication steps (lithography, etch step) and the post processing fabrication steps (dicing, polishing, anti-reflection (AR) coating). In a next step, the characterization of rib waveguides will be described. Measurements on straight and bent waveguides are presented. It follows a description of geometrical birefringence in SOI rib waveguides. Furthermore, a birefringence tuning method based on cladding induced stress is presented, which can be applied to keep the polarization dependent frequency shift (PDFS) of a MZ-DI as small as possible.

Chapter 4 describes basic theory and simulation of MMI-couplers, which are key elements of MZ-DIs. Then, the simulated characteristics (excess loss, imbalance, bandwidth) for specific geometries of a 2×2 and a 4×4 MMI coupler as well as optical measurements on these couplers are presented. It follows a consideration of MMI coupler fabrication tolerances.

Chapter 5 specifies the MZ-DI layout. It follows the presentation of 2×2 and 2×4 MZ-DI measurements. Finally, the impact of fabrication tolerances on PDFS will be investigated.

Chapter 6 and Chapter 7 present two different applications of the SOI platform. The first application is a demodulator for DPSK signals. The second application refers to an all-optical wavelength converter. For both applications, device requirements and chip layouts will be presented. BER performances are plotted. Related to the interaction between passive and active components, the hybrid integration of BPDs and SOAs will be investigated. Finally, the realization of device packages will be described.

Novel results in this PhD-thesis

- Demonstration of a high-performance SOI Mach-Zehnder delay interferometer for 40 Gb/s DPSK receiver applications, which meets the system specifications
- Demonstration of fully passive C-band high-performance optical 90°-hybrids in 4µm SOI technology
- Demonstration of a high-speed (40Gb/s) all-optical wavelength converter using semiconductor optical amplifiers flip-chip integrated on SOI motherboard with cascaded delay interferometers
- Demonstration of a high-precision flip-chip technology for III-V active components on 4µm SOI motherboards

2 Mach-Zehnder delay interferometer for phase demodulation

2.1 Modulation formats

Over many years optical communication systems were dominated by the use of conventional on-off keyed (OOK) signals, i.e. intensity modulated signals. In the recent past, a number of advanced formats (beyond OOK [9]) for high speed transmission applications have been studied. The information is carried by the optical phase (phase shift keying, PSK). More specifically, to avoid the need of an absolute phase reference, differential-PSK (DPSK) considers the optical phase change between adjacent bits. Compared to OOK, the balanced detection of DPSK signals provides a ~3 dB in terms of average optical power. This leads to higher receiver sensitivity or lower OSNR requirements and can be exploited to extend transmission distances, reduced optical power requirements or relaxed component specifications [10].

Fig. 2.1. shows signal diagrams of OOK, DPSK and DQPSK signals [9].The DPSK-format has a 0 or π phase-shift between adjacent bits. The DPSK format can be extended to differential quadrature phase-shift keying (DQPSK) with four phase levels, which are shifted by $\pi/2$. The levels at $\{0, \pi/2, \pi, 3\pi/2\}$ allow the encoding of 2 bits per symbol. This can be used either to double the data rate compared to DPSK based encoding while maintaining the bandwidth or to maintain the data rate of DPSK with half of the required bandwidth.

The 3 dB power benefit of DPSK signals compared to OOK signals can be intuitively understood by considering the amplitudes of the optical signal.

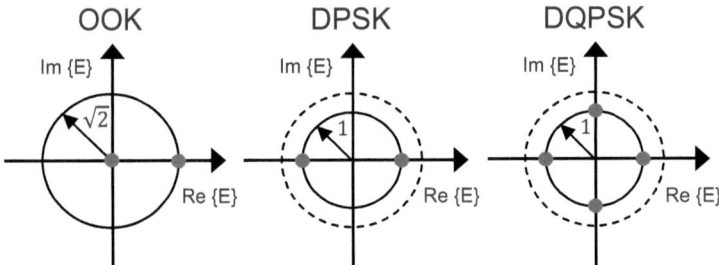

Fig. 2.1 Signal constellations of binary on-off keying (OOK), differential PSK and QPSK. The dashed line represents the optical field amplitude of OOK [10].

Obviously, only half of the average optical power is needed for DPSK signals as compared to OOK to obtain the same symbol distance [10].

2.2 General phase demodulator scheme

The demodulation of DPSK signals can be carried out by 2×2 Mach-Zehnder delay interferometers (MZ-DIs) as shown in Fig. 2.2. (a). The MZ-DI notation follows the number of in- and outgoing waveguides. The structure is composed by two 2×2 MMI couplers, a delay line and waveguides for in and out coupling. The delay line length ΔL determines a phase shift $\Delta \varphi$ given by $\beta \cdot \Delta L$, where β denotes the propagation constant of the optical mode. The first 2×2 MMI coupler in Fig. 2.2 (a) operates as a 3 dB splitter and ideally inserts a 90° phase shift on the input signal launched at input 1 (i1) or input 2 (i2).

The outgoing signals of the 2×2 MMI coupler are then delayed by one interferometer arm corresponding to 1 symbol period, which is equally to 1 bit period in case of the DPSK signals. The second 2×2 MMI coupler combines the signals of both interferometer arms, i.e. "compares" the phases of two consecutive symbols.

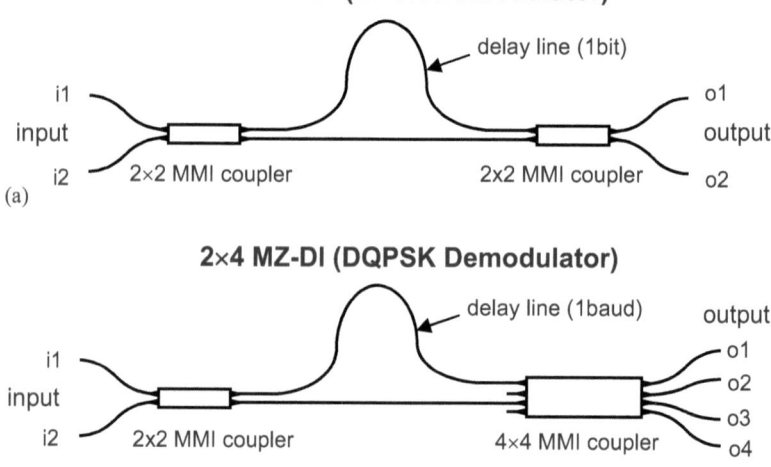

Fig. 2.2 Scheme of 2×2 (a) and 2×4 (b) Mach-Zehnder delay interferometer (MZ-DI) for phase demodulation of DPSK and DQPSK signals, respectively.

In case of equal phases of two consecutive symbols, the MZ-DI output signals are maximized at o2 due to constructive interference and minimized at output 1 due to destructive interference. The phase information is thus transformed into intensity information. Unequal phases of two consecutive bits lead to high signal at o1 and low signal at o2. The detection at the output ports can be carried out by so called single-ended detection or by use of balanced photo detectors (BPD). It has to be mentioned, that only the balanced detection yields the 3 dB power benefit.

MZ-DIs are also useful for demodulation of DQPSK signals. One implementation uses two MZ-DIs to execute the demodulation of each quadrature component. The incoming signal is split into two equal parts and passes the interferometers with phase difference of $\pi/2$. The interferometers have the same delay length, corresponding to the used symbol time and equal to twice of the bit duration. A more compact implementation is the use of a 90°-hybrid, which is included into a 2×4 MZ-DI. A scheme of this interferometer structure is shown in Fig. 2.2. (b). Compared to the DPSK demodulator, one 2×2 MMI coupler has to be replaced by an optical 90° hybrid, which can be implemented by a 4×4 MMI coupler. The requirements regarding the PDFS are still higher than for DPSK demodulators. The MZ-DI related requirements will be considered in more detail in the next section.

2.3 Ideal and Non-ideal MZ-DI – a description with matrix theory

The transfer function of a Mach-Zehnder delay interferometer can be described by basic matrix theory [11; 12]. Here, a 2×2 MZ-DI can be represented by a black box with two inputs and two outputs, see Fig. 2.3. The input and output E-field of the optical signal is denoted with $\underline{E}_{i,2x2}$ and $\underline{E}_{o,2x2}$, respectively. In the following, we shall limit our formalism to the forward transmission case, i.e. we shall neglect reflection terms.

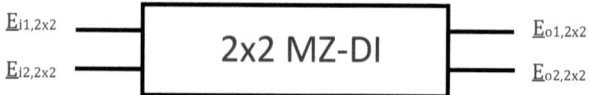

Fig. 2.3 Simplified 2×2 Mach-Zehnder delay interferometer structure for matrix description.

In general, the complete 2×2 MZ-DI transfer function can be written as:

$$\begin{pmatrix} \underline{E}_{o1,2x2} \\ \underline{E}_{o2,2x2} \end{pmatrix} = \underline{T}_{MZDI,2x2} \begin{pmatrix} \underline{E}_{i1,2x2} \\ \underline{E}_{i2,2x2} \end{pmatrix} \quad (2.1)$$

Here $\underline{T}_{MZDI,2x2}$ denotes the transfer matrix of the 2×2 Mach-Zehnder delay interferometer, which is obtained by the the product of partial matrices of the MMI coupler $\underline{T}_{MMI}$ and the phase shift section $\underline{T}_{\Delta\varphi,2x2}$:

$$\underline{T}_{MZDI,2x2} = \underline{T}_{MMI,2x2} \, \underline{T}_{\Delta\varphi,2x2} \, \underline{T}_{MMI,2x2} \quad (2.2)$$

With respect to different characteristics of the two MMI couplers, we can write the MMI coupler matrix elements as follows:

$$\underline{T}_{MMI,2x2} = \begin{pmatrix} a_i & jb_i \\ jb_i & a_i \end{pmatrix} \quad (2.3)$$

The characters a_i and b_i are real numbers, which describe the coupling coefficients of the i-th MMI coupler (i=1, 2) of the MZ-DI. The matrices for the description of lossless MMI couplers have to be unitary, i.e. $a_i^2 + b_i^2 = 1$. Also assumed are the symmetry and the reciprocity of the matrix elements.

The ideal 2×2 Mach-Zehnder delay interferometer

The ideal Mach-Zehnder delay interferometer assumes perfect MMI couplers (no loss, accurate 3dB-splitting, 90° phase shift between the outputs) and a lossless phase shift section. This behavior is valid for both MMI couplers of the MZ-DI. Hence, we can specify $a_i = b_i = 1/\sqrt{2}$ leading to the following transfer matrix of the 2×2 MMI coupler ($\underline{T}_{MMI,2x2}$):

$$\underline{T}_{MMI,2x2} = \frac{1}{\sqrt{2}} \begin{pmatrix} 1 & j \\ j & 1 \end{pmatrix} \quad (2.4)$$

The transfer matrix of the phase shift section ($\underline{T}_{\Delta\varphi}$) is given by:

$$\underline{T}_{\Delta\varphi} = \begin{pmatrix} e^{j\varphi_1} & 0 \\ 0 & e^{j\varphi_2} \end{pmatrix} = e^{j\varphi_2} \begin{pmatrix} e^{j\Delta\varphi} & 0 \\ 0 & 1 \end{pmatrix} \quad (2.5)$$

Here, $\Delta\varphi$ denotes the phase difference between the two interferometer arms and can be described with β, which is the propagation constant of the fundamental waveguide mode in the interferometer arms. Furthermore, the interferometer arms have the length L with an additional length ΔL (corresponding to a time delay of 1 symbol period) in one of the interferometer arm. This results in the following expression for $\Delta\varphi$:

$$\Delta\varphi = \varphi_1 - \varphi_2 = \beta(L + \Delta L) - \beta L = \beta \Delta L \quad (2.6)$$

By use of only one input ($\underline{E}_{i2,2x2}=0$), the complete transfer function can be written with eq.(2.2) as follows:

$$\begin{pmatrix} \underline{E}_{o1,2x2} \\ \underline{E}_{o2,2x2} \end{pmatrix} = \underline{T}_{MZDI,2x2} \begin{pmatrix} \underline{E}_{1,2x2} \\ 0 \end{pmatrix} = \underline{T}_{MMI,2x2} \underline{T}_{\Delta\varphi} \underline{T}_{MMI,2x2} \begin{pmatrix} \underline{E}_{1,2x2} \\ 0 \end{pmatrix} \quad (2.7)$$

$$\begin{pmatrix} \underline{E}_{o1,2x2} \\ \underline{E}_{o2,2x2} \end{pmatrix} = \frac{e^{j\varphi_2}}{\sqrt{2}} \begin{pmatrix} e^{j\Delta\varphi} - 1 \\ j(1 + e^{j\Delta\varphi}) \end{pmatrix} \underline{E}_{i1,2x2} \quad (2.8)$$

The optical intensity at the two MZ-DI output ports is given by:

$$\begin{pmatrix} P_{o1,2x2} \\ P_{o2,2x2} \end{pmatrix} = \frac{1}{2} \begin{pmatrix} 1 - \cos(\Delta\varphi) \\ 1 + \cos(\Delta\varphi) \end{pmatrix} |\underline{E}_{i1,2x2}|^2 = \begin{pmatrix} \sin^2(\frac{\Delta\varphi}{2}) \\ \cos^2(\frac{\Delta\varphi}{2}) \end{pmatrix} |\underline{E}_{i1,2x2}|^2 \quad (2.9)$$

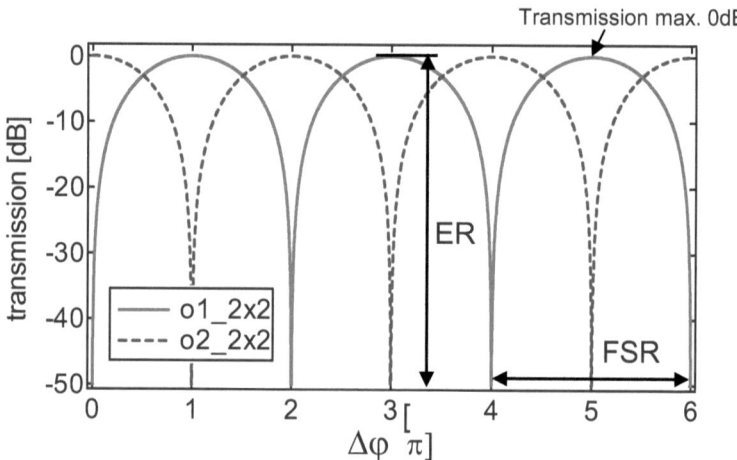

Fig. 2.4 Power transmissions (o1, o2) of an ideal 2×2 Mach-Zehnder delay interferometer vs. phase. The transmissions show no insertion loss (transmission maxima are at 0 dB). Ideal extinction ratio (ER) of -∞ is not shown beyond −50 dB.

Obviously, a phase shift of π enables switching of light from one port to the other port. This situation is shown in Fig. 2.4. Under the assumption of ideal power splitting and phase relationships we do not observe any reduction in the normalized transmission maxima of both ports. Therefore, insertion loss (IL) defined as the deviation in dB from the maximum available transmission (here at 0 dB) does not exist. Furthermore, the extinction ratio of an ideal MZ-DI, which is the ratio between maximal and minimal MZ-DI transmission in decibel is -∞. In Fig. 2.4 the transmission plot is not shown beyond -50 dB. Also denoted is the free-spectral range (FSR) of a MZ-DI, which corresponds to the distance of two transmission minima given by a phase shift of 2π.

The ideal 2×4 Mach-Zehnder delay interferometer
The transfer matrix of an ideal 2×4 MZ-DI can be described in a way similar to the ideal 2×2 MZ-DI described in the last section. Fig. 2.5 shows a simplified schematic of a 2×4 MZ-DI with the labeling for the inputs and outputs for matrix description.

Fig. 2.5 Simplified 2×4 MZ-DI structure for matrix description.

As shown in Fig. 2.6 (b), one 2×2 MMI coupler of the 2×2 MZ-DI is replaced by a 4×4 MMI coupler. The ideal transfer matrix of a 4×4 MMI coupler ($\underline{T}_{MMI,4x4}$) is given by Bachmann [13]:

$$\underline{T}_{MMI,4x4} = \frac{1}{\sqrt{2}} \begin{pmatrix} e^{j\pi} & e^{-j\frac{\pi}{4}} & e^{j\frac{3\pi}{4}} & e^{j\pi} \\ e^{-j\frac{\pi}{4}} & e^{j\pi} & e^{j\pi} & e^{j\frac{3\pi}{4}} \\ e^{j\frac{3\pi}{4}} & e^{j\pi} & e^{j\pi} & e^{j\frac{7\pi}{4}} \\ e^{j\pi} & e^{j\frac{3\pi}{4}} & e^{j\frac{7\pi}{4}} & e^{j\pi} \end{pmatrix} \quad (2.10)$$

The complete transfer function of the 2×4 MZ-DI can be calculated as follows:

$$\begin{pmatrix} \underline{E}_{o1,2x4} \\ \underline{E}_{o2,2x4} \\ \underline{E}_{o3,2x4} \\ \underline{E}_{o4,2x4} \end{pmatrix} = \underline{T}_{MMI,4x4} \, \underline{T}_{\Delta\varphi} \, \underline{T}_{MMI,2x2} \begin{pmatrix} \underline{E}_{i1,2x4} \\ 0 \\ \underline{E}_{i2,2x4} \\ 0 \end{pmatrix} \quad (2.11)$$

$$= \frac{e^{j\varphi_2}}{2\sqrt{2}} \begin{pmatrix} e^{j\pi} & e^{-j\frac{\pi}{4}} & e^{j\frac{3\pi}{4}} & e^{j\pi} \\ e^{-j\frac{\pi}{4}} & e^{j\pi} & e^{j\pi} & e^{j\frac{3\pi}{4}} \\ e^{j\frac{3\pi}{4}} & e^{j\pi} & e^{j\pi} & e^{j\frac{7\pi}{4}} \\ e^{j\pi} & e^{j\frac{3\pi}{4}} & e^{j\frac{7\pi}{4}} & e^{j\pi} \end{pmatrix} \begin{pmatrix} e^{j\Delta\varphi} & 0 & 0 & 0 \\ 0 & 0 & 0 & 0 \\ 0 & 0 & 1 & 0 \\ 0 & 0 & 0 & 0 \end{pmatrix} \begin{pmatrix} 1 & 0 & j & 0 \\ 0 & 0 & 0 & 0 \\ j & 0 & 1 & 0 \\ 0 & 0 & 0 & 0 \end{pmatrix} \begin{pmatrix} \underline{E}_{i1,2x4} \\ 0 \\ \underline{E}_{i2,2x4} \\ 0 \end{pmatrix}$$

The following expression results for the 2×4 MZ-DI output signals by use of the first input port:

$$\begin{pmatrix} \underline{E}_{o1,2x4} \\ \underline{E}_{o2,2x4} \\ \underline{E}_{o3,2x4} \\ \underline{E}_{o4,2x4} \end{pmatrix} = \frac{e^{j\varphi_2}}{2\sqrt{2}} \begin{pmatrix} e^{j\Delta\varphi}e^{j\pi} + je^{j\frac{3\pi}{4}} \\ e^{j\Delta\varphi}e^{-j\frac{\pi}{4}} + je^{j\pi} \\ e^{j\Delta\varphi}e^{j\frac{3\pi}{4}} + je^{j\pi} \\ e^{j\Delta\varphi}e^{j\pi} + je^{j\frac{7\pi}{4}} \end{pmatrix} \underline{E}_{i1,2x4} \quad (2.12)$$

Thus, we obtain the following expression for the output power of the 2×4 MZ-DI:

$$\begin{pmatrix} P_{o1,2x4} \\ P_{o2,2x4} \\ P_{o3,2x4} \\ P_{o4,2x4} \end{pmatrix} = \frac{1}{2} \begin{pmatrix} \cos^2(\Delta\varphi/2) \\ \sin^2(\Delta\varphi/2 - \pi/4) \\ \cos^2(\Delta\varphi/2 - \pi/4) \\ \sin^2(\Delta\varphi/2) \end{pmatrix} |\underline{E}_{i1,2x4}|^2 \quad (2.13)$$

The calculated output power of the 2×4 MZ-DI for $|\underline{E}_{i1,2x4}|^2 = 1$ is plotted in Fig. 2.7. The transmission signals of the output ports are shifted by $\pi/2$. The maximum output signal is – 3 dB.

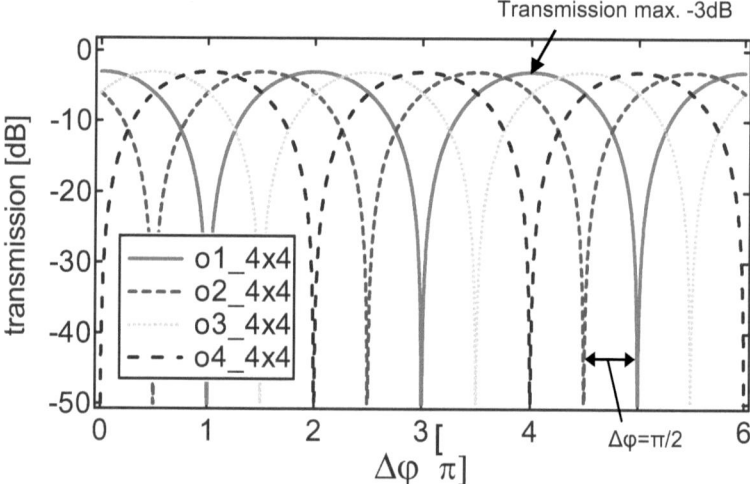

Fig. 2.7 Power transmission of an ideal 2×4 MZ-DI vs. phase. The phase difference between two transmission minima is $\pi/2$.

The non-ideal 2×2 Mach-Zehnder delay interferometer

A "non-ideal" MZ-DI shall be described in terms of factors, which affect the extinction ratio of the ideal MZ-DI [11; 12]. The extinction ratio is influenced by the imperfect MMI coupler (phase errors and unequal power splitting ratios), unbalanced losses in the interferometer arms as well as polarization dependence of MMI couplers and waveguides. The influence of imperfect power splitting ratios in a 2×2 MMI coupler and unbalanced interferometer arm losses are considered here by use of matrix formalism. For this, equation (2.5) for the description of the phase section has to be rewritten as follows:

$$\underline{T}_{\Delta\varphi,2x2} = \begin{pmatrix} \alpha e^{j\Delta\varphi} & 0 \\ 0 & 1 \end{pmatrix} \qquad (2.14)$$

The character α is a real number in the range of 0 to 1 and denotes a differential loss between the two interferometer arms. In this case the loss is related to the delayed interferometer arm. As an example, a value of $\alpha=1$ means "lossless" (no differential loss) and a value of $\alpha = 0$ means infinite loss, which converges towards infinity. Furthermore, a variation in the individual coupling coefficients of the MMI couplers is considered, i.e. $a_1 \neq a_2 \neq b_1 \neq b_2$. Here, a_1 and b_1 corresponds to the coupling coefficients of the first MMI coupler and a_2 and b_2 of the coupling coefficients of the second MMI coupler (see 2×2 MZ-DI in Fig. 2.2). Also, the MMI couplers are considered to be lossless ($a_1^2 + b_1^2 = 1$ and $a_2^2 + b_2^2 = 1$.

After calculation of $\underline{T}_{MZDI,2x2}$ we obtain the following expressions for the output power:

$$\begin{pmatrix} P_{o1,2x2} \\ P_{o2,2x2} \end{pmatrix} = \begin{pmatrix} a_1^2 a_2^2 \alpha^2 + b_1^2 b_2^2 - a_1 a_2 b_1 b_2 \alpha \, 2\cos(\Delta\varphi) \\ a_1^2 b_2^2 \alpha^2 + a_2^2 b_1^2 + a_1 a_2 b_1 b_2 \alpha \, 2\cos(\Delta\varphi) \end{pmatrix} |\underline{E}_{i1,2x2}|^2 \qquad (2.15)$$

The extinction ratios in decibel of the non-ideal MZ-DI can be calculated with (2.15). At first, the extinction ratio of the individual outputs is given by minimum and maximum available transmission:

$$ER_{o1,2x2} = 10\log\left(\frac{\max P_{o1,2x2}}{\min P_{o1,2x2}}\right) \qquad (2.16)$$

$$ER_{o2,2x2} = 10\log\left(\frac{\max P_{o2,2x2}}{\min P_{o2,2x2}}\right) \qquad (2.17)$$

Together with eq.(2.15) this leads to two different expressions for the achievable extinction ratios at the two MZ-DI outputs:

$$ER_{o1,2x2} = 10\log\frac{(a_1a_2\alpha + b_1b_2)^2}{(a_1a_2\alpha - b_1b_2)^2} \quad (2.18)$$

$$ER_{o2,2x2} = 10\log\frac{(a_1b_2\alpha + a_2b_1)^2}{(a_1b_2\alpha - a_2b_1)^2} \quad (2.19)$$

The output power vs. $\Delta\varphi$ is depicted in Fig. 2.8 for two constellations: In (a) a loss of 1 dB in one interferometer arm is assumed. This results in a clear extinction ratio reduction to about 25 dB. The curves in (b) arise from a coupler imbalance of $a_i^2/b_i^2 = $ 55%/45 % in both MMI couplers, respectively. Here, the interferometer arms are assumed to be lossless. As a consequence of the coupler imbalance, the extinction ratios at both MZ-DI outputs differ. At parallel output port (o1) only 20 dB is reached while extinction at cross port remains unchanged.

Equations (2.18) and (2.19) are plotted in Fig. 2.9 for different interferometer arm losses and MMI coupler splitting ratios. As shown in Fig. 2.8, the extinction ratio degrades at both output ports with increased loss in one of the interferometer arms (see the extended plot in Fig. 2.9 a).

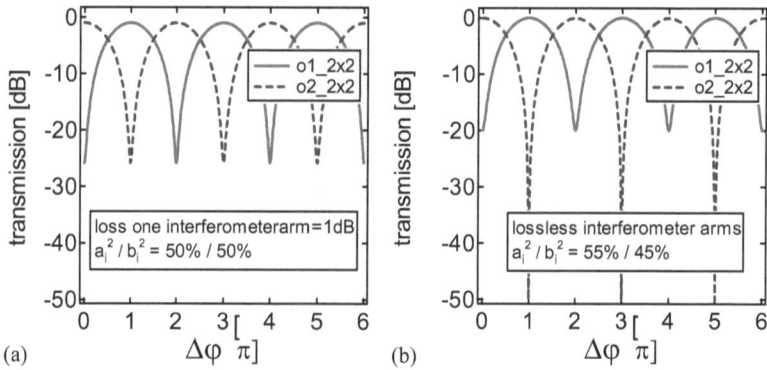

Fig. 2.8 Transmission spectra of non-ideal MZ-DIs resulting from equation (2.15). In (a), an inserted loss of 1dB (corresponding to $\alpha = 0.891$) in one delay arm leads to clear reduction in extinction ratio at both ports. In (b), the MMI coupler imbalance of $a_i^2/b_i^2=0.55/0.45$ degrades clearly the extinction ratio at the parallel output port (o1). In (b), the interferometer arms are assumed to be lossless, i.e. $\alpha=1$.

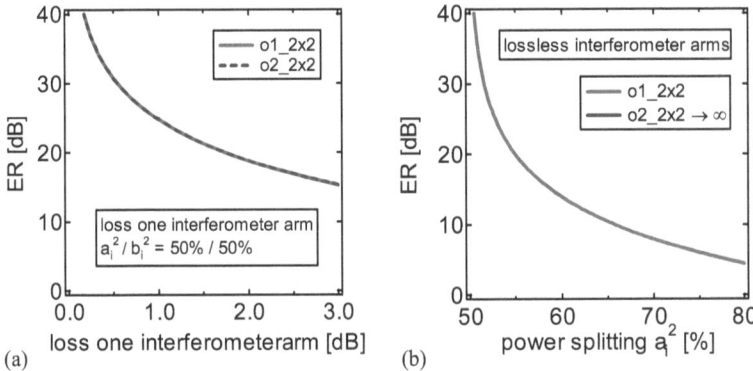

Fig. 2.9 Extinction ratio reduction due to loss in one interferometer arm (a) Here, both MMI couplers of the MZ-DI have perfect power splitting. In (b), the power splitting of both 2×2 MMI couplers is imperfect. In (b), the interferometer arms are assumed to be lossless.

Fortunately, losses for SOI rib waveguides with H = 4µm are typically in the range of 0.2 dB / cm. Therefore, the interferometer arm with an additional delay length introduces a significant loss only over long distances (~ cm), which would correspond to a very small free spectral range (< 8 GHz). In the case of 40 Gb/s DPSK demodulators (FSR = 40 GHz), the arm length difference is about 2 mm and leads therefore to a small additional loss in one interferometer arm. On the other hand, substrates with higher waveguide losses – such as photonic wires (typically 2 dB / cm) – suffer more from such constraints.

Fig. 2.9 (b) shows the influence of improper power splitting at the MMI couplers on the ER. For 2×2 MMI coupler with imbalances higher than 55% to 45% (> 0.9 dB), the ER in the parallel port (o1) degrades to values below 20 dB.

Nevertheless, the ideal MZ-DI extinction ratio of -∞ can be obtained in spite of delay line loss and MMI coupler imbalance. The ideal extinction ratio remains in case of a balance between delay line loss and MMI coupler imbalance, i.e. $a_1\alpha = b_1$ and $a_2 = b_2 = 1/\sqrt{2}$ (see equations (2.18) and (2.19)). However, the implementation of this situation is impaired by the difficult adjustment of variable MMI coupler imbalance.

Finally, the polarization dependence of MZ-DIs shall be considered. Polarization dependence is given by polarization-dependent loss or as polarization-dependent frequency shift (PDFS) induced by MZ-DI birefringence. The PDFS appears as a shift in the output transmission spectra.

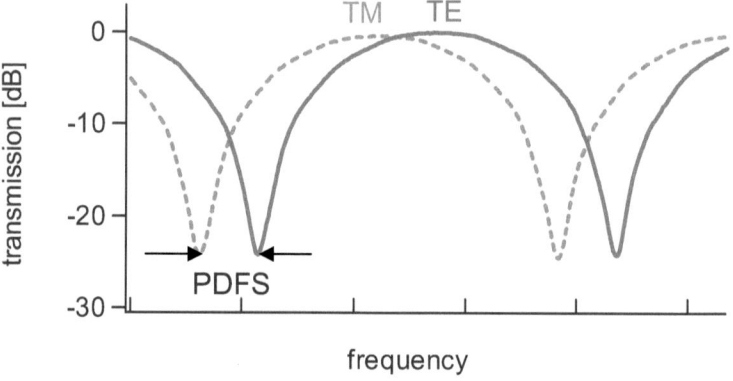

Fig. 2.10 Illustration of polarization-dependent frequency shift (PDFS) arising in transmission spectra of Mach-Zehnder delay interferometer with transverse electric (TE) and transverse magnetic (TM) light.

In is dependent on the state-of-polarization (SOP) of the input signal. As shown in [14], the PDFS can be described as a frequency offsets relative to a transmit laser frequency. Fig. 2.10 illustrates the effect of PDFS on MZI transmission spectra by using transverse electric (TE) and transverse magnetic (TM) polarized light.

In free space, the expressions "TE" and "TM" light describe the orientation of the electric and magnetic field oscillations with respect to the plane of incidence. In a planar waveguide, the electric field parallel to the wafer plane can be defined as TE light. The light polarized perpendicular to the wafer plane is defined as TM light. Light with arbitrary polarizations can then be composed of a linear superposition of TE and TM light.

2.4 Demodulator requirements

The effect of polarization dependent frequency shift on the performance of DPSK and DQPSK-demodulation was properly analyzed by Kim & Winzer [14; 15] and Bosco [16]. Calculations show a linear scaling of the tolerable PDFS with the bit rate. Frequency offsets between 4 to 5 % of DPSK bit rate leads to a sensitivity penalty of 1 dB. The corresponding PDFS is 1.6 to 2 GHz. The DQPSK format shows a ~ 6 times lower tolerance to frequency offsets compared to DPSK [14].

The following table (Tab. 2.1) shows the experimental results for the polarization dependence of 40 Gb/s DPSK and DQPSK systems of [14]. Here the allowed frequency offset for 1 dB sensitivity penalty of DPSK-system is only 3 % of the bit rate, i.e. 1.2 GHz in 40 Gb/s systems.

Tab. 2.1 Frequency offsets for sensitivity penalty of 1 dB and 2 dB in 40 Gb/s systems [14]. The experimental data are given in percent of the bit rate and as polarization dependent frequency shift (PDFS), respectively.

Sensitivity penalty	DPSK		DQPSK	
	% bit rate	PDFS	% bit rate	PDFS
1 dB	± 3%	± 1.2GHz	± 0.5%	± 0.2GHz
2 dB	± 5%	± 2GHz	± 0.8%	± 0.3GHz

We may calculate the required phase accuracy of the MMI couplers based on the tolerable PDFS (corresponding to 1 dB OSNR penalty as shown in Tab. 2.2). For 40 Gb/s DPSK systems it has to be better than ± 10° (= tolerable PDFS [GHz] · 360° / FSR [GHz]). For DQPSK receivers the phase accuracy has to be better than ± 2°.

Successful operation of the MZ-DIs for DPSK and DQPSK demodulation has been demonstrated over the past years in fiber- and free space- technology. Planar lightwave circuit (PLC) developments mainly focused on silica waveguide-technology. In different technologies a PDFS of $\leq$ 1GHz could be realized.

Table Tab. 2.3 summarizes the important results for state-of-the-art performance in the area of demodulator technology implemented in different technologies. As can be seen from the table, Mach-Zehnder delay interferometers of the different technologies provide extinction ratios of 20 - 30 dB. The insertion loss of the fiber-based MZ-DIs is very low due to the low-loss fiber and the absence of lossy coupling interfaces. The sizes of silica-based PLC's (at higher complexity) show the potential for smaller devices. The next chapters will investigate whether SOI-technology can achieve such performance levels or even open a way for improvements.

Tab. 2.3 Performance of Mach-Zehnder delay interferometers for DPSK and DQPSK demodulation in different technologies. The table compares polarization dependent frequency shift (PDFS), extinction ratios (ER) and insertion loss (IL). The insertion loss characterizes the additional loss of the signal power by insertion of an optical device into a transmission line. Also shown is the measured optical signal-to-noise ratio (OSNR) at a bit-error rate (BER) of 10^{-9}.

Design/ Published	Format/ Bitrate	Size Volume or Area	PDFS [GHz]	ER [dB]	IL [dB]	OSNR [dB] at BER = 10^{-9}
Free-space optical Michelson-Interferometer/ Liu et al [17], Dec. 2005	DPSK RZ/NRZ 42.7 Gb/s	7.3cm^3	0.3	25	1.5	no OSNR rec. power -36.3dBm
All-Fiber/ Lizè et al [18], Dec. 2007	DPSK/ 43 Gb/s	23.8cm^3	0.5	30	0.3	-
Silica-on-silicon/ Doerr et al [19], Jan. 2006	DQPSK RZ 42.7 Gb/s	0.38 cm^2	1	20	4.5 - 5.8	21.5
Silica-on-silicon/ Oguma et al [20], Sept.2007	DQPSK/ 43 Gb/s	0.52 cm^2	0.1	25	4.3 - 5	20.1

3   4µm SOI rib waveguides

3.1   SOI material system & single-mode condition

SOI material has a silicon layer of a height H with relatively high refractive index n_{Si} of ~ 3.47, which can be used as wave-guiding system. An oxide layer with refractive index n_{SiO2} of ~ 1.44 is used for the lower cladding, while air or an oxide layer is used as the top cladding layer. The high refractive index difference between the core and the cladding enables a high confinement of the waveguide modes. A waveguide design can be carried out as rib-waveguide. The rib waveguide can be realized by only one etching step, which ensures optical confinement in the lateral direction. This is shown in Fig. 3.1.

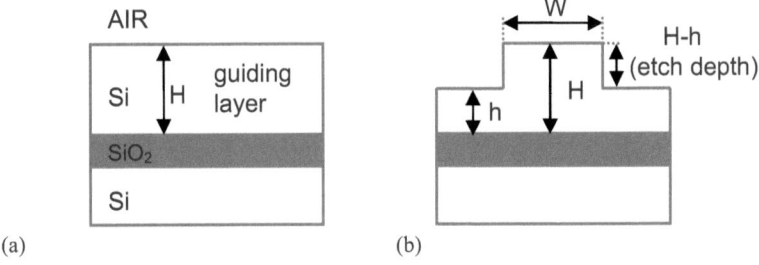

Fig. 3.1 The three layer material system of the SOI-technology (a) and the scheme of a rib waveguide (b). The rib waveguide dimensions can be described by the rib waveguide width W, the rib height H and the slab height h.

Although high index contrast waveguides with large cross section support many modes, SOI rib waveguides can be designed to enable only single-mode (SM) operation. That requirement is important for devices like interferometers to avoid performance degradation due to dispersion effects. Petermann and Soref presented conditions for single-mode case of rib waveguides [21; 22]. Under the assumption of $r > 0.5$ and relatively large cross sections $\left(H \geq \lambda / \sqrt{n_{Si}^2 - n_{SiO2}^2}\right)$ [21] single-mode can be achieved by proper choice of the rib height H, the slab height h and width W of the rib leading to an approximation for single-mode case given by:

$$\frac{W}{H} \leq c + \frac{r}{\sqrt{1-r^2}} \qquad (3.1)$$

The ratio r is given by the ratio of lateral cladding thickness h to the rib height H:

$$r = \frac{h}{H} \qquad (3.2)$$

The constant c amounts to 0.3. The character λ corresponds to the free-space optical wavelength. In the vertical direction higher order modes of the rib waveguide are cut off by coupling into modes of the slab area. That happens if the effective index of the higher order modes in the rib region becomes lower than the effective index of the fundamental mode in the slab area.

Fig. 3.2 illustrates Soref's single-mode condition by use of a normalized rib width (t = W / H) and a normalized slab height (r = h / H). The rib height H is fixed at 4µm. The single-mode condition of Soref was derived only for the case of r > 0.5. For other waveguide geometries (0.3 ≤ r ≤ 0.8) another single-mode condition has been proposed by Aalto [23], which is shown in the same figure. Aalto's single mode condition is spitted into two domains with a crossing point of both domains at about r = 0.43. Fig. 3.2 shows also the calculated SM condition by use of a commercial mode solver FIMMWAVE[1] by PhotonDesign. The results are in good agreement with the Petermann/Soref results for r > 0.5.

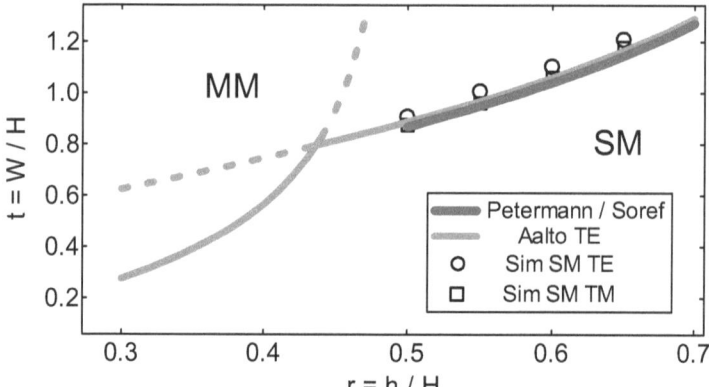

Fig. 3.2 Single-mode condition of Petermann/Soref and Aalto (TE mode) for 4 µm SOI waveguides using normalized rib width and normalized slab height. Aalto's single-mode mode condition for TM mode is very similar and therefore not presented here. The calculated SM condition by FIMMWAVE is plotted by circles (TE mode) and rectangles (TM mode).

---

[1] The FIMMWAVE tool of PhotonDesign provides a fully-vectorial solver based on the film-mode matching (FMM) method for calculation of 2D-mode profiles and related propagation constants.

Generally, single mode operation at r < 0.5 implies reduced waveguide width, which results in higher waveguide loss (see Rickman [24], Zinke [25]). Therefore, the range of r ≥ 0.5 has more practical relevance. For example, for a rib waveguide with H = 4 µm and a slab thickness of h = 2 µm, the single-mode condition will be fulfilled for a waveguide width ≤ 3.6 µm.

3.2 Optical Losses in SOI rib waveguides

The intrinsic optical loss in low-doped silicon is low for wavelengths > 1.2 µm. That yields very low-loss rib waveguides (~ 0.1 dB / cm) as shown by Fischer [7]. However, depending on lithography and etching steps, fabricated rib waveguides can suffer considerably from scattering loss. Scattering loss in SOI waveguides results from surface roughness σ. A deeper analysis of waveguide losses was done by Payne and Lacey [26]. They found an upper bound for scattering loss proportional to σ^2 and $1/d^4$, where d denotes the half width of the guiding layer (d = H/2). This shows the high influence of surface roughness and waveguide dimensions on waveguide loss. A reduction of the surface roughness after waveguide processing might be possible by thermal oxidation.

Perturbations of the rib waveguide structure due to mask errors, resist problems or etch inhomogeneities may also cause mode conversion resulting in leaky modes. Notwithstanding the manifold ways of introducing scattering loss, the relevance of losses in 4µm SOI waveguide technology is rather low.

Bending loss

Bends require a careful design to avoid radiation loss by mode coupling from the rib waveguide into the cladding material. The key design parameter in that context is the bending radius R. This has been demonstrated by the analysis of waveguide bends by Marcatili and Miller. They have shown that the waveguide loss decreases exponentially with increasing bending radii [27].

Fig.3.3 shows the calculated TE mode power distribution for SOI rib geometry of H = 4.0 µm, W = 3.5 µm and r = 0.5 for two different bending radii. For calculation, a bend mode solver by PhotonDesign using complex finite difference method (FDM) was deployed. Obviously, the mode power extends into the slab at smaller bending radii (R = 1.5 mm). At R = 5 mm the mode shape seems to be little influenced by the bend.

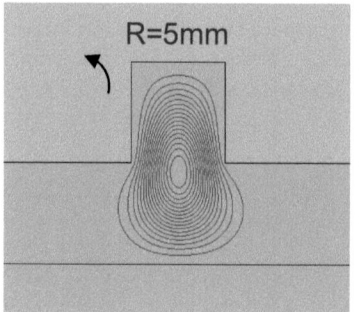 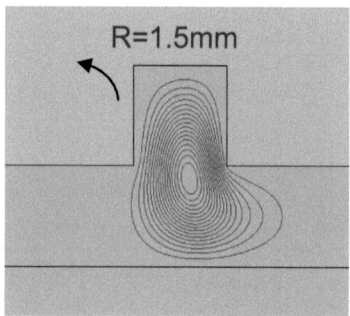

Fig.3.3 TE-mode power distribution of (left oriented) bends in SOI rib waveguides (H = 4.0 μm, W = 3.5 μm, r = 0.5, λ = 1.55 μm) at R = 5 mm and R = 1.5 mm.

To keep the radiation loss of the propagating mode as low as possible, the waveguide bend has to be relatively large. To estimate the loss of a bend, a further analysis was carried out with the complex bend mode solver of PhotonDesign (λ = 1550 nm). The result is plotted in Fig. 3.4. For 90° bends the radius is varied between 1 mm to 6 mm. In the region of small radii we get clearly higher losses for the TE-polarized light. Obviously, the horizontal confinement of the TM mode is stronger than that of the TE mode. For a radius R of 1.5 mm the loss amounts for TE mode amounts to 6.7 dB and for TM mode to 1.1 dB. Such high and additionally polarization dependent loss is undesirable in photonic devices. The bending loss for TE and TM mode at R = 4 mm is less than 0.1 dB.

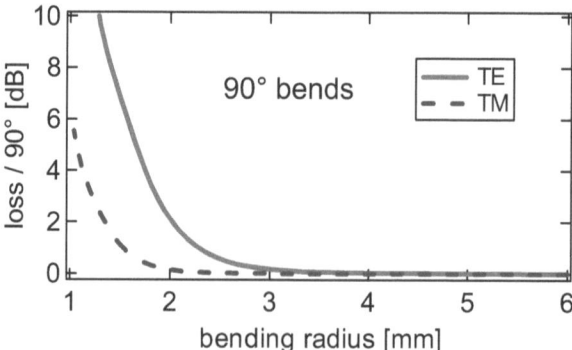

Fig. 3.4 Calculated loss of 90° bends as a function of the rib waveguide bending radius for TE and TM mode (H = 4 μm, W = 3.5 μm, r = 0.5, λ = 1.55 μm).

The simulated bending loss is small for both polarizations when the bending radius is more than 4 mm. Loss for both polarizations is less than 0.1 dB for bending radius larger than 4 mm. However, the calculation only indicates a design rule. A radius of 4 - 5 mm seems to be suitable for low loss 90° bends in 4µm SOI technology. Additional scattering loss in bends given by increased side wall roughness through process imperfections is not considered here. Experimental results of straight and bent SOI waveguides will be presented in section 3.6.

A common approach for the reduction of bending loss in low index-contrast waveguide systems could be the introduction of waveguide offsets at the beginning and at the end of a waveguide bend [28]. Calculation of the offset in our 4µm rib waveguide technology show that the offset of the mode profile increases with smaller bend radius, see Fig. 3.5. At a bend radius of 5 mm an offset of 0.12 µm would be required. The offsets are determined by the lateral peak shift of the rib waveguide mode profile as function of the bend radius.

The mode profiles of the bent rib waveguides were calculated by use of the FDM Solver in FIMMWAVE (PhotonDesign). Since waveguide offsets in ~ 100 nm range are beyond the fabrication capabilities of our lithographic tools, we decided to design bends without offsets.

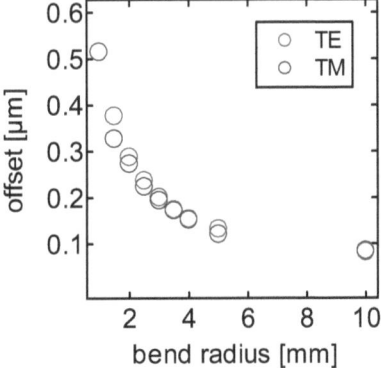

Fig. 3.5 Calculated rib waveguide offsets as function of bend radius for TE and TM mode.

3.3 Coupling single-mode fiber (SMF) - SOI rib waveguide

The mode-field diameter (MFD) of a standard single-mode fiber differs from the MFD of a rib waveguide in 4µm SOI technology by a factor of ~ 2. Therefore, we may expect a loss by coupling of the modal fields of a standard single-mode fiber and a rib waveguide (H = 4 µm, W = 3.5 µm, H-h = 2 µm). The coupling loss can be determined by calculation of the overlap between the modal fields with the FIMMPROP tool of PhotonDesign. Calculations indicate a coupling loss of about 6 dB, which is too big for optical applications. Therefore, a spot-size conversion is required. The conversion is not lossless and can be carried out on-chip or off-chip. Tab. 3.1 shows the pros and cons:

Tab. 3.1 Pros and Cons for on-chip and off-ship spot-size conversion.

Approach	Advantage	Disadvantage
on-chip, i.e. 3D-taper	match SMF	topography fluctuations, extra processing effort
off-chip, i.e. lensed fiber	no additional processing	alignment tolerances

Throughout this thesis, fiber-chip coupling is achieved by means of lensed fibers. This approach is beneficial due to considerably lower processing efforts, avoiding epitaxial growth and topography fluctuations on the wafer. However, implications for fiber pigtailing are small alignment tolerances, which are a significant drawback for packaging.

Fig. 3.6 shows calculated horizontal (x) and vertical (y) alignment tolerances for the system with lensed fiber (MFD ~ 3.3µm) and SOI rib waveguide (H = 4 µm, W = 3.5 µm, H-h = 2 µm). This calculation is based on the consideration of overlap between a misaligned lensed fiber and a rib waveguide. The coupling loss increases strongly in x and y direction. A lateral or vertical displacement of 0.5 µm leads to an additional coupling loss of ~0.6 dB. A displacement of 1 µm leads to an additional coupling loss of ~2 dB. During the calculation of misalignment tolerances, the distance between lensed fiber and rib waveguide is zero (z = 0 µm). Any pigtailing technology suffers from misalignment in both directions, x and y.

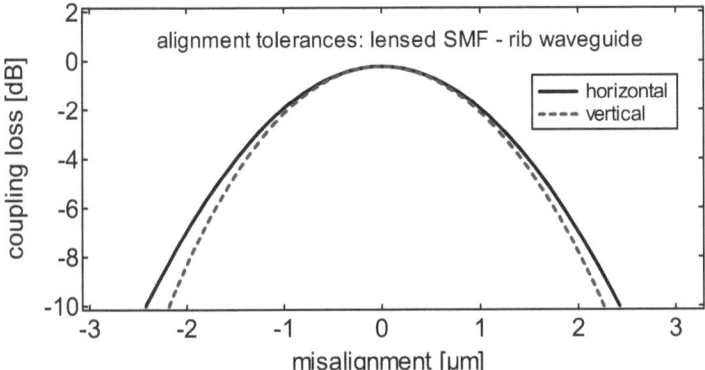

Fig. 3.6 Alignment tolerances in horizontal and vertical direction for the coupling of lensed single-mode fiber (SMF) – rib waveguide (H = 4μm, W = 3.5μm, H-h = 2μm). The distance z between lensed fiber and rib waveguide during the calculations is zero (z = 0 μm). The minimum coupling loss is ~ 0.2 dB.

Therefore, even if pigtailing can be done with a fiber placement accuracy of 0.5μm, losses of misalignment in x as well as in y direction will add up to ~1.2 dB power penalty. Fiber pigtailing of such small spot-size devices is highly demanding with regard to placement and fixation of the fiber in x and y direction. A variation in z-direction (distance between lensed fiber and rib waveguide) is not considered here. At z = 0 and optimum alignment (misalignment x, y = 0), the calculated coupling loss between lensed fiber and rib waveguide amounts to 0.2 dB.

3.4 Fabrication of SOI rib waveguides

The fabrication process of SOI rib waveguides is straightforward. Usually, only one lithographic and one etch step are sufficient to generate a waveguide structure. However, the successful realization of designed optical devices implies accurate knowledge of fabrication tools and technologies. In addition to that some precaution needs to be taken because planar waveguides are very susceptible to damage by inappropriate handling.

Post processing is necessary to ensure the measurability of fabricated optical devices. For the case of 4μm-SOI technology this requires the preparation of waveguide facets for optical coupling to external networks. The post processing consists of separation into single chips, polishing and anti-reflection coating (ARC) at the waveguide facets.

The fabrication starts with the bare SOI wafer. SOI wafers are fabricated in various ways, e.g. separation by implantation of oxygen (SIMOX), bonded and etched back silicon-on-insulator (BESOI), and smart-cut technologies. For the purposes of this PhD thesis primarily BESOI material was deployed.

The basic waveguide fabrication process includes the rib waveguide patterning and a rib waveguide etch. The waveguide patterning was carried out by use of standard contact photolithography (Süss, MJB 55 UV500). The required mask for transfer into positive resist was manufactured at Fraunhofer Institut für Nachrichtentechnik-HHI Berlin by e-beam writing.

The rib waveguide etch was done with in-house reactive-ion etching (Oxford, Plasmalab 80) based on fluorine chemistry (SF_6). The dry-etch process of silicon requires low temperatures (200 K). It offers the required anisotropic etch profiles and leads to low surface roughness. The selectivity between the used resist (AZ 5214) and silicon etching is sufficient ($\sim$ 1:4). Fig. 3.7 shows exemplary scanning electron microscope (SEM) pictures of dry-etched 2×2 and 4×4 MMI couplers in SOI material.[2] Following the silicon etch, the remaining resist was removed with wet chemicals and plasma oxidation. The clean room fabrication is completed by the deposition of a protection layer of silicon oxide with a thickness of about 500 nm.

The post processing starts with separation of the fabricated wafer into single chips by means of dicing. Afterwards follows a smoothing of the rib waveguide facets by a grinding and polishing procedure. The grinding and polishing for purposes of this thesis was done with a CL-40 from Struers (Fig. 3.8 a). Objective of the initial two-stage grinding with SiC-paper is the creation of a plane surface with only scratching damages. These damages can be mostly removed by chemo-mechanical polishing using specific polishing slurry (SF1) on a rotating polyurethane-disc.

For protection of the chip surface during polishing it is beneficial to deposit an additional layer of SiO_2 (500 nm) on chip surface. Helpful is also the use of a glass dummy on top of the chip. The glass dummy has approximately the same size as the SOI chip and is glued to the surface of the SOI by means of a thermoplastic adhesive. The "sandwich" (SOI-chip + glass dummy) stabilizes the waveguide facets during grinding and polishing, and protects them from mechanical damages.

The thermoplastic adhesive can be resolved with acetone after the polishing process. An additional cleaning of the chip surface and waveguide facets is furthermore possible by application of hydrofluoric acid and a plasma oxidation ($\sim$1 h).

---

[2] Special thanks to B. Maul (HHI Fraunhofer Berlin), who made the SEM pictures in this section.

Fig. 3.7 Scanning electron microscope picture of a 2×2 (a) and a 4×4 MMI-coupler (b) in 4μm SOI technology after reactive-ion etch (RIE) step.

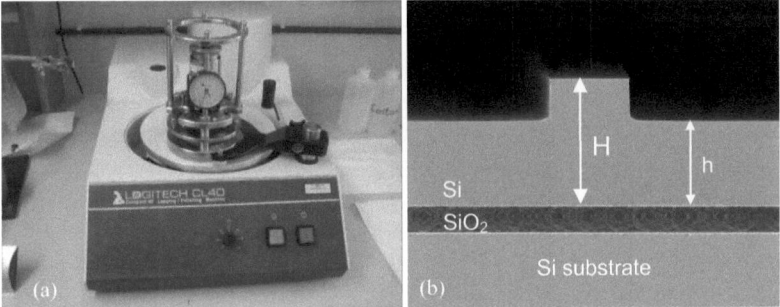

Fig. 3.8 Polishing machine CL-40 from Struers (a). SEM picture of a polished rib waveguide facet in 4μm SOI technology (b).

The interface air-silicon leads to reflections and therefore coupling loss of. At 1.5μm, reflection loss is ~1.5 dB (~ 30 % of input power). Therefore, the waveguide facets require an additionally ARC to keep optical losses based on reflections as low as possible. As shown by Schnarrenberger [29], the ARC can be realized by use of a silicon nitride layer with a refractive index of 1.86 and a thickness of 208 nm. Then the reflections are reduced down to 10^{-3} [29]. A SEM picture image of a polished rib waveguide after RIE etching and stripping is shown in Fig. 3.8 (b). The image shows high verticality and smoothness of the side-wall and the polished facet, respectively.

3.5 Measurement setup

The general optical characterization of 4μm SOI waveguides is based on transmission measurements, which enable the specification of:

- insertion loss (including coupling loss and waveguide loss)
- filter characteristics (imbalance, extinction ratios)
- bandwidth limitations
- polarization dependence

The measurement setup consists of a tunable laser source (TLS), a polarization controller and a calibration unit as well as a power sensor for signal recording. Fig. 3.9 shows a scheme of the setup used for optical measurements in the optical lab at TU Berlin.

The optical signal of a tunable laser source (Agilent 81940A) is injected into a programmable polarization controller (Agilent 8169A), which provides well defined states of polarization. The connection is achieved by use of standard single mode fibers (SMF). The chip rests on a temperature controlled stage.

Light is coupled in- and out by means of lensed fibers. Corning SM fibers with a mode-field diameter of ~ 3.3 μm are deployed in the setup. Finally, an optical power sensor (Agilent 81634B) records transmission characteristics. To obtain TE- and TM polarized light at the chip-input, a polarization calibration is required. The calibration setup is sketched in Fig. 3.9 (box as dashed line).

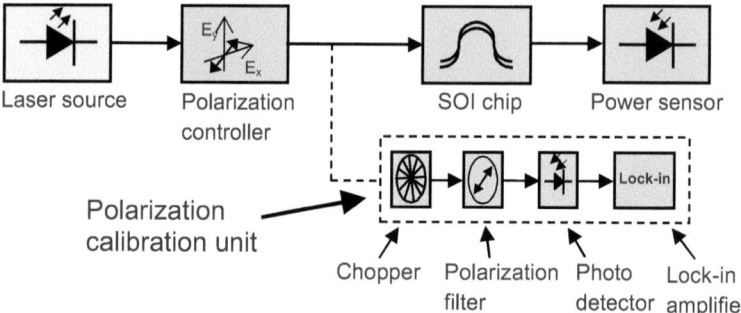

Fig. 3.9 Measurement setup for characterization of 4μm SOI waveguides. The connections are realized by SMF-fibers. The depicted free-space polarization calibration includes a chopper, a polarization filter, a photo detector and a lock-in amplifier for signal detection.

The signal of the incoming lensed fiber passes initially a polarization filter. It follows a shutter, which is connected to a lock-in amplifier to separate light signal from external signals. Well defined linear polarized light can now be adjusted by change in the ε & θ - settings of the polarization controller. In addition to linearly polarized light, the polarization controller allows for generation of random-polarized light, which covers the whole Poincaré sphere.

Fig. 3.10 shows typical signal levels in the measurement setup. The power of the optical light source is 6 dBm. Our back-to-back measurement includes a polarization controller and standard fiber connections, which already leads to a signal loss of about 1.7 dB. The fiber-to-fiber (F2F) measurement corresponds to a transmission line consisting of a polarization controller, standard fibers and two lensed fibers. The lensed fibers are adjusted to each other to reach a maximum optical signal in the transmission line. The F2F measurements include additionally a signal loss given by the spot-size conversion from standard SMF to lensed SMF. This conversion loss is about 1 dB / lensed fiber, i.e. 2 dB by use of two lensed fibers.

The signal of a reference waveguide measurement is still less (minimum 1 dB) than the F2F measurement signal due to coupling loss between lensed fiber and standard 4µm-SOI rib waveguide with anti-reflection coating (0.5 dB / facet) plus rib waveguide loss (≥ 0.1 dB / cm). The SOI reference waveguide consists of a straight waveguide with s-bend at about half of the total waveguide length.

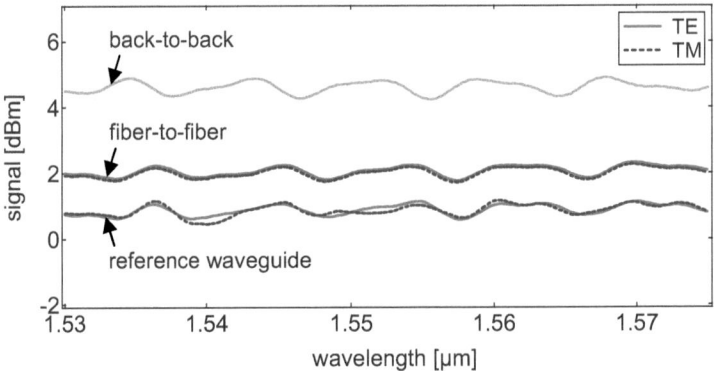

Fig. 3.10 Signals of the measurement system at TU-Berlin. The plot shows signals of a back-to-back (B2B) measurement, a fiber-to-fiber (F2F) measurement and of a reference waveguide.

All transmission measurements in this thesis are plotted with respect to reference waveguide measurements. This means, the transmission graphs show directly excess loss and additional waveguide loss (caused by increased waveguide length) of the device under test. Measurements of reference waveguides itself refer to the F2F-signal.

3.6 Characterization of straight and bent waveguides

For regular monitoring of waveguide losses each chip design at TUB contains straight waveguides, which include an s-bend section. The use of s-bends ensures that only guided light is detected at the output facet. Fig. 3.11 (a) shows results of an optical transmission measurement of straight waveguides. 20 waveguides were fabricated under nominal conditions (H = 4 µm, H-h ≤ 2 µm, W = 3.5 µm), then polished & coated with an antireflection layer. The intrinsic loss is very uniform. The normalised waveguide losses are in the range between 0.1 dB / cm and 0.2 dB / cm. Polarization dependent loss (PDL) is not significant in 4 µm SOI waveguides.

It is also required to estimate a minimum radius that ensures low intrinsic waveguide losses together with minimum polarization dependence. The minimum bending radius can be determined by plotting waveguide loss as a function of waveguide bending radius. Fig. 3.11 (b) collects the results of transmission loss measurements of bends with an angle of 90° (H = 4 µm, H-h ≤ 2 µm, W = 3.5 µm) with varying radius.

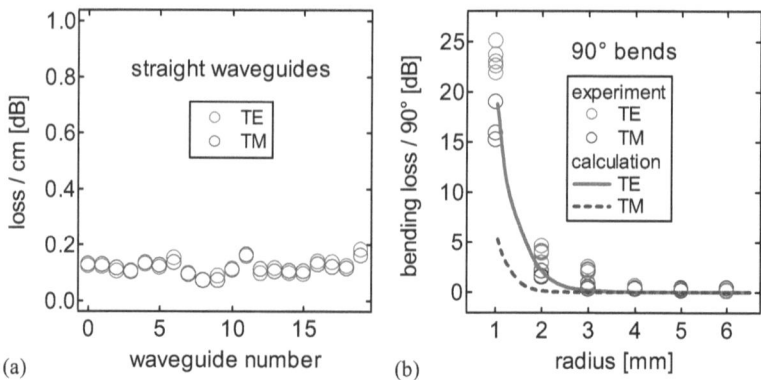

Fig. 3.11 Normalized intrinsic waveguide losses of straight waveguides (a). Loss of 90°-bends with different bending radii (b). Lines indicate loss calculations, circles the measured loss values.

Beginning at R = 6 mm, the losses increase slightly for both polarizations at R = 4 mm. The additional losses at R = 2 mm are higher than 1 dB for TM and almost 4 dB for TE polarization, respectively. To overcome process variations, the minimum bending radius in waveguide layouts should therefore not be less than ~ 5 mm. Despite a slight offset, the measurements o R ≥ 4 mm are in good agreement with the calculated curves (see Fig. 3.4). For R < 4 mm, the difference between calculated and measured curves increases.

3.7 Polarization dependence of SOI rib waveguides

One of the biggest challenges in planar waveguide technology is polarization independent waveguide performance. Optical fibers feed optical signals with arbitrary polarization states to the waveguides. Therefore arbitrary states of polarizations have to be processed on chip while maintaining high performance. The polarization dependence can appear as birefringence and polarization dependent loss (PDL). The latter is evoked by sidewall roughness, causing different losses for TE and TM polarized modes. In interferometers both effects can occur and may lead to performance degradation [14; 15], even more if they are present at the same time.

Waveguides with squared cross section show ideally no birefringence. In waveguides with different aspect ratios (W/H) there is modal birefringence, which is usually not negligible. Modal waveguide birefringence as function of TE and TM modes can be defined as follows:

$$\Delta n_{eff} = n_{eff}^{TE} - n_{eff}^{TM} \qquad (3.3)$$

Here n_{eff}^{TE} and n_{eff}^{TM} denote the effective refractive indices of the TE and TM modes. The effective refractive indices are defined by the propagation constants of the fundamental TE and TM mode (β_0^{TE}, β_0^{TM}) in a rib waveguide, i.e. $n_{eff}^{TE} = \beta_0^{TE}/k_0$ and $n_{eff}^{TM} = \beta_0^{TM}/k_0$ with free-space wave number $k_0 = 2\pi / \lambda$. Besides that modal birefringence, we also have to consider stress induced birefringence Δn_{stress}. The overall waveguide birefringence is therefore given by:

$$\Delta n_{eff} = \Delta n_{geo} + \Delta n_{stress} \qquad (3.4)$$

As shown in Fig. 2.10, the birefringence leads to a polarization dependent frequency shift (PDFS) in a spectral MZ-DI response. The PDFS can be calculated by considering the optical wavelengths of the two fundamental modes.

The wavelengths are equal under assumption of the following relation:

$$\lambda_{TE}(f_1) = \lambda_{TM}(f_2) \tag{3.5}$$

The frequencies f_1 and f_2 describe the PDFS (equivalent to $2\Delta f$) with respect to the frequency of light f, i.e. $f_1 = f - \Delta f$ and $f_2 = f + \Delta f$. The wavelengths for TE and TM mode in equation (3.5) can be calculated as follows:

$$\frac{c}{(f - \Delta f)\, n_{eff}^{TE}(f - \Delta f)} = \frac{c}{(f + \Delta f)\, n_{eff}^{TM}(f + \Delta f)} \tag{3.6}$$

This expression can be written as:

$$(f - \Delta f)\left[n_{eff}^{TE}(f) - \frac{dn_{eff}^{TE}}{df}\Delta f \right] = (f + \Delta f)\left[n_{eff}^{TM}(f) + \frac{dn_{eff}^{TM}}{df}\Delta f \right] \tag{3.7}$$

Finally, the PDFS can be defined as follows:

$$\text{PDFS} = 2\Delta f \approx \Delta n_{eff}\, f/\overline{n_g} \tag{3.8}$$

Here, $\overline{n_g}$ describes the averaged group index of refraction for TE and TM mode, i.e. $(n_g^{TE} + n_g^{TM})/2$. The group index is defined by $n_g^{TE,TM} = n_{eff}^{TE,TM} + f\, dn_{eff}^{TE,TM}/df$. The group indices for TE and TM mode can be calculated by use of a mode solver (FIMMWAVE, PhotonDesign). For standard rib waveguide dimensions (H = 4 µm, H-h = 2 µm, W = 3.5 µm) result group indices as follows: $n_{g;TE} = 3.608929$, $n_{g;TM} = 3.60932$ ($\lambda = 1550$ nm). The PDFS in interferometric devices for DPSK demodulation should not exceed 1 GHz (see Tab. 2.1). This limits the birefringence in SOI waveguides to $2\cdot10^{-5}$. DQPSK receiver systems tighten the requirement on birefringence to about $3\cdot10^{-6}$ [14].

3.7.1 Geometrical birefringence

Fig. 3.12 shows the dependence of modal birefringence on r and W values. The simulations were carried out with the FIMMWAVE tool of PhotonDesign. As already shown for 5µm SOI rib waveguides [30], we can reduce the birefringence of 4 µm rib waveguides by modification of the rib waveguide geometry.

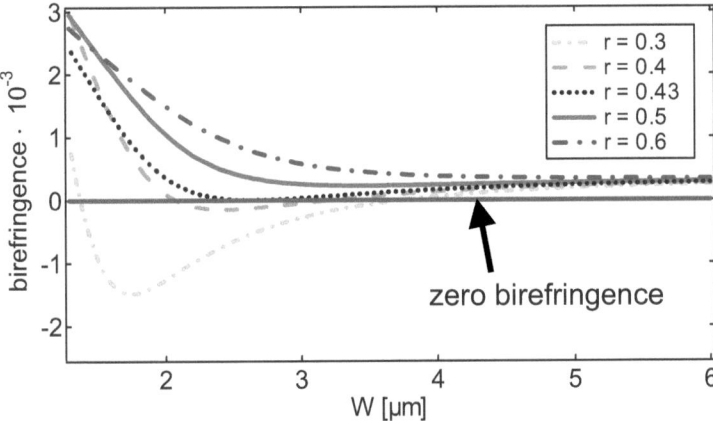

Fig. 3.12 Birefringence as function of rib waveguide geometry (H = 4µm). For r = 0.43 and smaller r values zero birefringence is achievable.

At r = 0.43 and a rib width W of about 2.7 µm, the birefringence drops to zero. The zero birefringence for r ≤ 0.43 arises at two different rib widths, which diverge with smaller value of r. Zero birefringence waveguides should be single-mode. This is analyzed in Fig. 3.13 as function of the normalized parameters t (W/H) and r (h/H). As shown in Fig. 3.12, zero birefringence occurs only for r ≤ 0.43. In terms of Aalto's single-mode condition, only the small range of r between 0.4 and 0.43 for zero birefringence would ensure single-mode operation. The strong dependence of zero birefringence on exactly defined rib waveguide geometry is a big challenge for waveguide fabrication. The close proximity to multimode operation is an additional challenge. Therefore, a purely geometrical approach to zero birefringence operation does not seem to be appropriate.

The SOI thickness has also an effect on waveguide birefringence. So far, the SOI thickness was assumed to be exactly 4 µm. Real wafer material shows thickness non-uniformities. Simulated birefringence data for three different SOI thicknesses (±5 and ±10 % of ideal H) are shown in Fig. 3.14 (a). The rib waveguide width W varies between 1 and 6 µm. We obtain for varying W in all substrates significant effects in birefringence. However, the depicted curves coincide at a waveguide width of about 3.5 µm very closely and make the birefringence almost independent from variations in the SOI substrate.

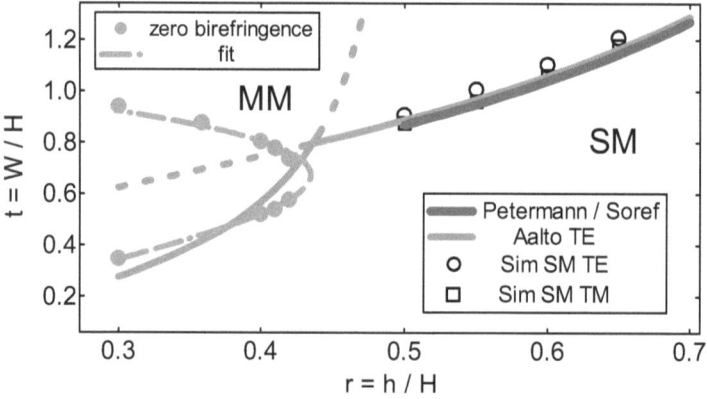

Fig. 3.13 Non-birefringence as function of the normalized parameters t (=W/H) and r (=h/H). Here a large cross section of the SOI rib waveguide is assumed (H = 4 µm).

In Fig. 3.14 (b) we vary the etch depths in different substrate types. The rib waveguide width W is constant 3.5 µm. The etch depth (H-h) varies between 1 and 3 µm. Generally, the birefringence decreases with larger (H-h) in all substrate types. On the other hand, the birefringence diverges between the substrates for etch depths substantially different from 2 µm.

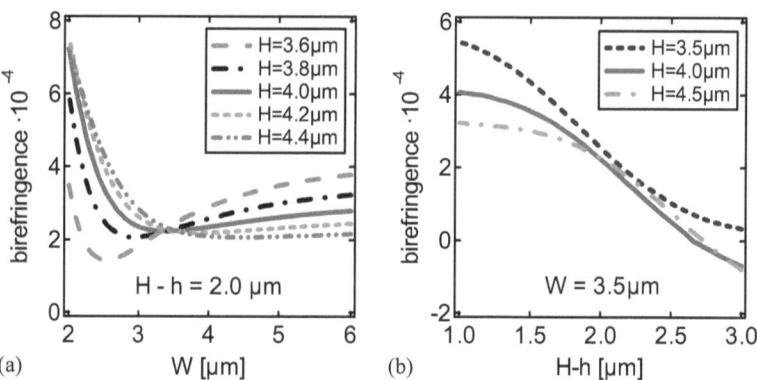

Fig. 3.14 Rib waveguide birefringence as function of substrate thickness. In (a) the rib width W varies between 1 and 6 µm. In (b) the etch depth varies from 1 to 3µm at W=3.5µm for three substrates.

If we want to avoid birefringence variations due to substrate or process non-uniformities, we should use a waveguide geometry around H = 4µm, W ~ 3.5µm and (H-h) = 2 µm. However, such SOI rib waveguides have considerable modal birefringence (~ 2.3·10^{-4}). The presented results show the large influence of waveguide geometry on birefringence. The birefringence can also be controlled by applied stress, which will be considered in the following section.

3.7.2 Stress induced birefringence

It has been shown in the section before that the intrinsic geometrical birefringence in the 4 µm SOI rib waveguide technology can be minimized by a proper height-to-width ratio. However, this approach is connected to very small fabrication tolerances. As shown in Huang and Xu [31; 32], the control of birefringence is also possible by deposition of dielectric layers on top of the waveguide structures (cladding). Anisotropic stress will e.g. be generated by the mismatch of the thermal expansion between the cladding and the waveguide material. This technique is still applicable after device fabrication has been completed.

Cladding materials like silicon dioxide are convenient because of the fact that they are often deposited on waveguide structures as protection layer. Therefore, the same layer can be used for two purposes. Using the photoelastic effect, the birefringence can effectively be controlled by the modification of the inherent cladding stress or by the cladding thickness. This offers the possibility of realizing birefringence-free rib waveguides with an initial $\Delta n_{geo} > 0$. Compared to the birefringence modification by the rib geometry, the birefringence can be fine-tuned even after wafer fabrication, e.g. by post-processing of single waveguide tiles. The following equations describe the birefringence change [32]:

$$n_x - n_0 = -C_1 \sigma_x - C_2 (\sigma_y + \sigma_z) \qquad (3.9)$$

$$n_y - n_0 = -C_1 \sigma_y - C_2 (\sigma_z + \sigma_x) \qquad (3.10)$$

Here, C_1 and C_2 denote stress-optic constants (Si: C_1=-17.79·10^{-12} 1/Pa, C_2=5.63·10^{-12} 1/Pa). Furthermore, n_x and n_y refer to the material refractive indices in x (horizontal) and y (vertical) direction; σ_x, σ_y and σ_z are the components of the stress tensor.

Silicon nitride cladding layers have strain with typical values of 100 MPa – 2 GPa leading to increase of the rib waveguide birefringence change by first order approximation in the area of 10^{-4}...10^{-3}.

This birefringence change corresponds to the geometric birefringence values typical for 4μm SOI rib waveguides. The increased birefringence can be utilized to shift MZ-DI transmissions of TE and TM polarized light in a way that the shift matches exactly one free spectral range (FSR), which is equivalent to a zero birefringence filter characteristic.

An experimental transmission of a MZ-DI for 40Gb/s DPSK signal demodulation before and after the birefringence tuning is plotted in Fig. 3.15. The PDFS is expressed as polarization dependent wavelength (PDλ) shift after: PDFS·λ^2/c.

The birefringence tuning reduces the PDλ-shift between the transmission minima from ~110 pm to less than 8 pm (1 GHz), which is desired for this receiver type. Such shift values are very small, therefore the question arises of how accurate measurements can be carried out.

To measure such small shifts, a measurement setup has to provide well defined linearly polarized light (TE, TM) as well as high wavelength resolution (~ 1pm). Furthermore it is necessary to control the temperature of the device. A PDλ-shift of 1 pm corresponds to a birefringence of $2.5·10^{-6}$. With $\Delta n = T\ dn/dT$ we should therefore be able to adjust temperatures with an accuracy of ~ 0.013K. Otherwise polarization effects can not clearly be separated from temperature effects.

The measurement setup at TU-Berlin (see Fig. 3.9) is able to perform wavelength scans with increments of 1 pm. Therefore, also small PDλ-shift can be determined.

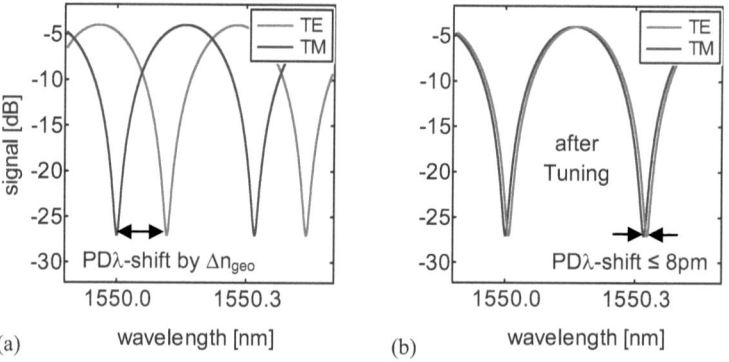

Fig. 3.15 Effect of birefringence tuning on transmission spectra of a MZ-DI for 40 Gb/s DPSK signal demodulation (FSR = 0.32 nm) in 4μm SOI technology. In (a) the polarization dependent wavelength shift (PDλ-shift) is given only by geometrical rib waveguide birefringence ($2.8·10^{-4}$). In (b) this birefringence is increased due to a stress-inducing cladding and can lead to a relative PDλ-shift of less than 8 pm (1 GHz).

However, minimal temperature variations on chip or fluctuations in the laser source during measurements can superimpose additional shifts. For that reason, single measurement points are not suitable for PDλ-shift determination. It is appropriate to estimate as many λ-shift values as possible. This statistic approach is visualized in Fig. 3.16. The upper part shows wavelength dependent MZ-DI transmission minima of TE and TM polarized light. The single PDλ-shift values are indicated with $\Delta\lambda_1$, $\Delta\lambda_2$, ..., $\Delta\lambda_i$. The exact values are plotted in the lower part of Fig. 3.16. and may have different algebraic signs.

The average of the PDλ-shift can be calculated by use of the root-mean-square (rms) function. Here, the rms function is defined by applying the square root on the arithmetic average of i single PDλ-shift values, i.e. $PD\lambda - \text{shift} = \sqrt{\frac{1}{i}(\Delta\lambda_1^2 + \Delta\lambda_2^2 + \cdots \Delta\lambda_i^2)}$. The experimental PDλ-shift or PDFS (8 pm ↔ 1 GHz) in this thesis refers to the rms-value.

Silicon oxide as well as silicon nitride with various thicknesses were tested at TU Berlin to estimate the birefringence change due to stress generation. The claddings were applied on the surface of Mach-Zehnder delay interferometers.

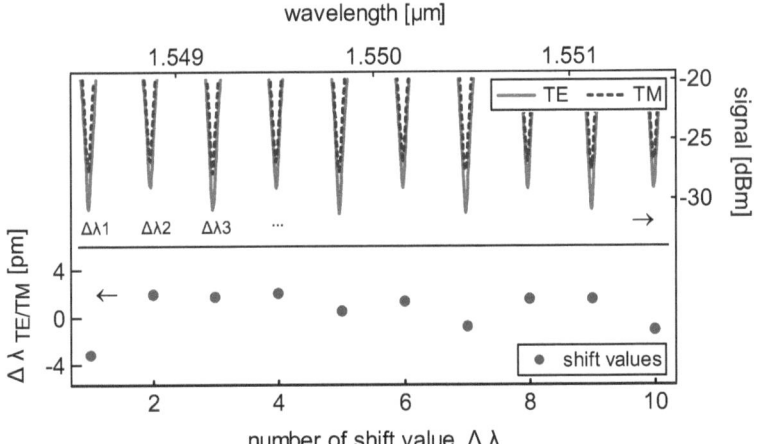

Fig. 3.16 Statistic approach for PDλ-shift estimation. On top: MZ-DI transmission minima of TE- and TM polarized light. Bottom: Single PDλ-shift values of the transmission minima in picometre (pm) scale.

Experimental results for birefringence change Δn_{stress} and corresponding polarization dependent frequency shift ($PDFS_{stress}$) are shown in Fig. 3.17 (a, b). The stress-induced birefringence Δn_{stress} was calculated from the polarization dependent wavelength shift $PD\lambda_{stress}$ in the transmission curves of TE and TM polarized light. Birefringence, polarization dependent wavelength and frequency shift are coupled as follows:

$$\frac{\Delta n_{stress}}{n_g} \approx -\frac{PD\lambda_{stress}}{\lambda} = \frac{PDFS_{stress}}{f} \quad (3.11)$$

Here, n_g denotes the group index of a waveguide (~ 3.6 in 4µm SOI rib waveguide) and λ and f the wavelength and frequency of light, respectively.

Silicon nitride Fig. 3.17 (a) and silicon oxide Fig. 3.17 (b) show opposite sign of stress induced birefringence by increase of cladding layer thickness. Due to the negative sign silicon oxide enables the reduction of total birefringence Δn_{eff} to zero.

A measured geometrical birefringence Δn_{geo} of ~ $2.8 \cdot 10^{-4}$ (device without cladding) can therefore be compensated with an oxide thickness of ~730 nm. However, the applied oxide degrades clearly the MZ-DI performance. The experiments showed increased insertion loss and decreased extinction ratios. The reason for this behavior could so far not be identified. The optical characteristics of silicon oxide might not involve this degradation.

Such performance degradation was not observed with silicon nitride cladding. The dependence between nitride thickness and stress induced positive birefringence can be linearly approximated (see Fig. 3.17 a). A nitride thickness $t_{Si_xN_y}$ of 205 nm leads to an additional birefringence of $5.2 \cdot 10^{-4}$ (or a PDFS of ~ 26 GHz). Summing PDFS of geometrical birefringence $PDFS_{geo}$ = 14 GHz ($\Delta n_{geo}=2.8 \cdot 10^{-4}$) and stress induced $PDFS_{stress}$ = 26 GHz ($\Delta n_{geo}=5.2 \cdot 10^{-4}$) leads to a separation of the transmission curves of TE and TM polarized light with a distance of exactly 40 GHz. The filter curves of an MZI realized in SOI technology for 40 GHz DPSK-demodulation can therefore be tuned to small polarization dependence, i.e. small relative PDFS.

It is also possible to combine the two cladding approaches. Here, the first cladding consists of silicon nitride and the second one of oxide. Thus, we can obtain two characteristics: high performance and total birefringence reduction. By application of this technique, the birefringence could be reduced to $4 \cdot 10^{-5}$ (corresponding to ~ 2 GHz PDFS). The cladding thicknesses were 205 nm silicon nitride and 4.3 µm silicon oxide, respectively. Here, no degradation of MZ-DI performance could be observed.

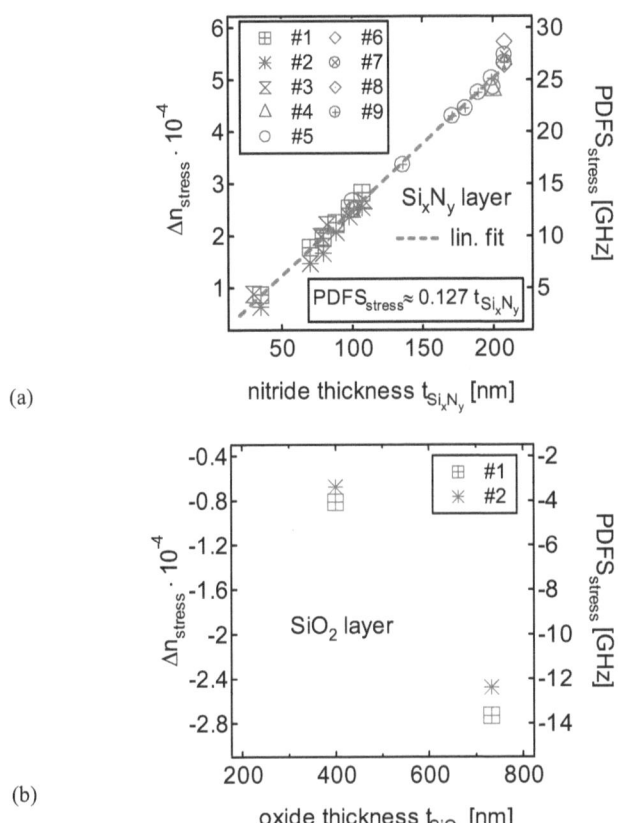

Fig. 3.17 Stress-induced birefringence (Δn_{stress}) and polarization dependent frequency shift (PDFS$_{stress}$) as function of cladding layer thickness t. The cladding material in (a) is silicon nitride (Si$_x$N$_y$), in (b) silicon oxide (SiO$_2$). Here "#" denotes the device number. The results base on the determination of PDFS on MZ-DIs with applied claddings. The depositions were carried out at process temperatures of 208°C (Si$_x$N$_y$) and 400°C (SiO$_2$).

However, the higher technology effort is a considerable disadvantage of this approach. The stress-induced birefringence tuning technique based on silicon nitride shows excellent applicability in 4µm SOI rib waveguide technology. So far, the birefringence tuning at TU-Berlin has been based on a standard silicon nitride deposition at a temperature of 208°C, which is also in use as anti-reflection layer.

The stress-induced birefringence can be controlled by the silicon nitride thickness. The generation of stress-induced birefringence at varying process conditions was so far not considered. It might be an option, to change the stress-induced birefringence by post-process thermal treatment as shown in [33]. Here, the oxide film stress increases with treatment temperature and time.

Birefringence tuning is also possible using half wave plates as shown for silica material in [19; 34; 20]. Such plates are inserted in the centre of MZ-DI arms to convert TE polarized light into TM light and vice versa. The TE and TM polarized light passes through the same optical path lengths, thus resulting in high interference quality.

However, the assembly of the wave plate implies extra effort with regard to the dicing of slots and bonding of the wave plate. Reflections caused by the plates may increase the insertion loss of devices.

Another approach uses a polarization diversity strategy. The separated TE and TM light will be handled in two different circuits, each optimised to process one particular polarization. In this approach, polarization splitters and combiners have to fulfil high standards related to splitting/combining performance. Another drawback is the required doubling of the processing units, which increases the footprint of the device and necessitates additional tuning.

State of the art is also the use of stress-releasing grooves. This is demonstrated in silica technology for a DQPSK demodulator [35]. By proper choice of path lengths with and without grooves, a PDFS close to zero can be achieved (~ 0.2 GHz). Moreover, this technique offers high potential for a reduction of heater power used for MZ-DI tuning.

4 Multi-mode interference (MMI) coupler

4.1 Design of MMI couplers

Basic components of the proposed MZ-DIs for phase demodulation (section 2.2) are multi-mode interference (MMI) couplers. They are useful because of their large variety of possible power splitting and combining characteristics. Additionally, MMI coupler phase relations can be utilized to obtain phase dependent switching. Compared to directional couplers, Y-branches or star couplers, MMI couplers in SOI technology prove to be the most robust splitter devices. The fabrication tolerances are relaxed. The underlying principle of the self-imaging was presented in the 70's by Bryngdahl [36] and Ulrich [37]. A practical approach is given by Soldano and Pennings [38]. The self-imaging principle can be described as a property of the multimode waveguides, which reproduces an input field profile periodically along the propagation direction of a multimode waveguide. The reproduced field images appear as single or multiple images.

The general MMI coupler structure consists of a central rib waveguide section, which is embedded in a system of access rib waveguides. For a good MMI coupler functionality, the central section should guide a sufficient number of modes. This number is not specified by theory. Based on a large number of 2×2 MMI coupler performance tests in 4µm SOI material, the number of guided modes should be a minimum of 8-9.[3]

The access waveguides are arranged at the beginning and the end of the central waveguide. The access waveguides provide at input waveguides a launch field (fundamental mode), which excites higher order modes in the central waveguide section. As output waveguides, they collect power from the multimode waveguide section. The notation N×N corresponds to N input and output waveguides. A schematic of a 2×2 and 4×4 MMI coupler is shown in Fig. 4.1. The MMI couplers have a characteristic length L_{mmi}, width W_{mmi} and access waveguide width W_{in}. The distance between MMI coupler edge and the centre of the access width is given by the character x. The in- and output waveguides are denoted by "i" and "o", respectively.

---

[3] The tests were carried out with the FIMMWAVE and FIMMPROP tools by PhotonDesign. The number of guided modes in the central section of the MMI coupler influence the energy transfer between access waveguide and central MMI coupler section. However, the MMI coupler performance results also from a number of other parameters as will be shown in the following sections.

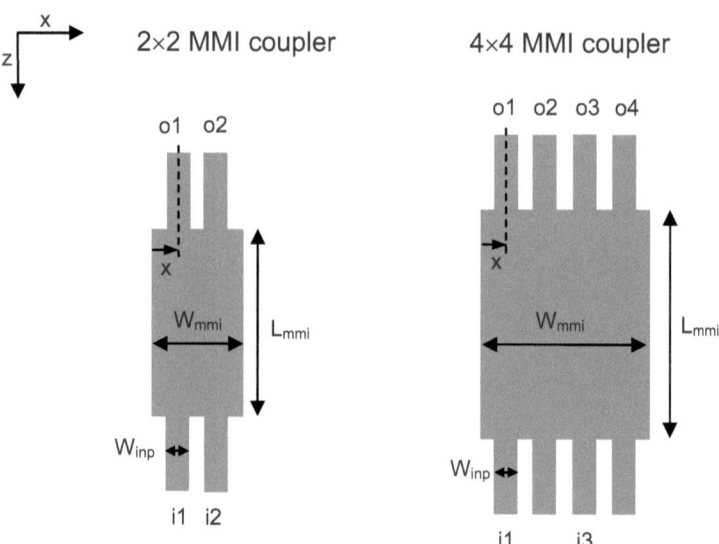

Fig. 4.1 Schematic of a 2×2 and a 4×4 multimode interference (MMI) coupler with length L_{mmi}, width of MMI coupler W_{mmi} and input waveguides W_{inp}. The character x denotes the distance between MMI coupler edge and access waveguide centre.

The realization of MZ-DIs for phase demodulation requires an optimum design of MMI couplers. A design procedure has to quantify the following MMI coupler characteristics:

- excess loss (EL)
- port imbalance (IMB)
- phase accuracy

The excess loss describes the relation between the input power and the output power of the MMI couplers. The excess loss of a 2×2 MMI coupler can be expressed as follows:

$$\text{EL(dB)} = -10\log\left(\frac{P_{o1} + P_{o2}}{P_{ref}}\right) \qquad (4.1)$$

Here, the output power of the MMI coupler is denoted by P_{o1} and P_{o2}. The input power, usually applied on one of the input waveguides (i1 or i2), is given by P_{ref}.

The imbalance can be defined by:

$$\text{IMB(dB)} = -10\log\left(\frac{P_{o1}}{P_{o2}}\right) \quad (4.2)$$

In case of 4×4 MMI couplers, equation (4.1) has to account for the output power of the four output ports. The used input waveguide has to be specified because of the possible impact on excess loss. The imbalance can be defined as required. For example, the balanced detection of demodulated DQPSK-signals requires a low imbalance between 4×4 MMI coupler output ports o2 and o3 as well as o1 and o4. Excess loss and imbalance have to be kept as low as possible and ideally they have to be polarization independent.

The design of MMI couplers is straightforward according to the basic theory described in [38]. However, no general approach is provided, which yields a MMI coupler with optimum performance. Therefore, the influence of various parameters on MMI coupler characteristics needs to be investigated. In this work, the MMI coupler optimization is focused on the following geometrical parameters:

- width and length of the MMI coupler
- position and width of the in- and output waveguides

The influence of these parameters will be studied in the following subsections using the FIMMWAVE and FIMMPROP tools by PhotonDesign. The first tool calculates the distribution and the propagation constants of the modes in the MMI coupler sections (access waveguides, multimode waveguide). The second tool calculates the overlap integral between the coupler sections. The device optimization can now be done by a suitable variation of the geometrical parameters mentioned above.

MMI coupler width and length

Generally, the length of a MMI coupler scales quadratically with MMI coupler width ($L_{mmi} \sim W_{mmi}^2$). The specific MMI coupler length depends on the desired output configuration, i.e. number of in- and output images, and the position of access waveguides relative to the central MMI coupler section. As shown in Tab. 4.1, the MMI coupler length can then be calculated in units of the beat length. The beat length L_π denotes the position of the first single image in z-direction (propagation direction) and can be calculated by the propagation constants of the fundamental and the first order mode ($ß_0$ and $ß_1$) in the central MMI coupler section, i.e. by $\pi / (ß_0 - ß_1)$.

Tab. 4.1 Examples of MMI couplers and their length at N-fold image at a given input waveguide position on the central MMI section.

MMI coupler (1×N, N×N)	position input waveguide $x / W_{mmi,eff}$	L_{mmi} at N-fold image
1×2	1/2	3/8 L_π
1×4	1/2	3/16 L_π
2×2	1/3	1/2 L_π
4×4	1/8, 3/8	3/4 L_π

The effective MMI coupler width ($W_{mmi,eff}$) reflects the fact that the optical modes of a rib waveguide penetrates into the surrounding material of a rib waveguide. Therefore, the effective MMI coupler width is slightly larger than the geometrical MMI coupler width. The effective width of planar waveguides can be calculated with an equation by Schnarrenberger [29]:

$$W_{mmi,eff} = W_{mmi} - 0.3H + \left(\frac{\lambda}{\pi}\right) \cdot \left(\frac{n_c}{n_r}\right)^{2\vartheta} \cdot (n_r^2 - n_c^2)^{-\frac{1}{2}} \quad (4.3)$$

This equation is modified by an additional term (0.3H) compared with the expression for the effective width of a planar dielectric optical waveguide as presented by Ulrich [37] and reflects the different mode field distributions of planar waveguides and rib waveguides. In equation (4.3), the effective refractive indices of a planar waveguide and the surrounding cladding are denoted by n_r and n_c, respectively. The exponent ϑ in eq. (4.3) is $\vartheta = 0$ for TE polarization and $\vartheta = 1$ for TM polarization. The effective refractive indices can be calculated by the propagation constants of the fundamental modes (β_0) in planar SOI waveguides, which have the rib height H and the slab height h (as defined for rib waveguide geometry), i.e. $n_r = \beta_{0,height\_H}/k_0$ and $n_c = \beta_{0,height\_h}/k_0$ with free-space wave number $k_0 = 2\pi / \lambda$.

The MMI coupler length dependence on the input waveguide position can be explained by different excitation of the modes in the central section of a MMI. Fig. 4.2 shows the power splitting of the fundamental TM mode of a rib waveguide (H = 4.0 µm, H-h = 2.0 µm, W_{inp} = 5 µm) into the TM modes of the MMI coupler multimode section in SOI rib waveguide technology (H = 4.0 µm, H-h = 2.0 µm, W_{mmi} = 30 µm). Here, the position of the input waveguide moves from the edge to the multimode waveguide centre. The resulting mode power distribution in the multimode rib waveguide is plotted versus effective width, which is slightly higher than the geometrical width W_{mmi} = 30 µm.

The splitting of the input waveguide power (fundamental mode) into modes of the central MMI coupler depends significantly on the position of the input waveguide.

Two specific situations can be observed: a positioning in the centre of the MMI coupler ($W_{mmi,eff}/2$) excites only symmetric or even modes (m=0, 2, 4, 6, ...). This characteristic can be utilized to realize a 1×N MMI coupler. The position at $W_{mmi,eff}/3$ (and $2W_{mmi,eff}/3$) of the effective MMI coupler width leads to pairwise excitation of modes (0-1, 3-4, 6-7, ...) - also referred to as restricted interference. The modes with index m = 2, 5, 8, ... are only weakly excited. This leads to a length reduction of the 2×2 MMI couplers compared to 2×2 MMI couplers using general interference mechanism [38]. The investigation of 2×2 MMI couplers in this thesis is focused on devices with restricted interference.

In the general interference case, the position of access waveguides is not predetermined. Here, a larger number of modes will be excited compared to restricted interference mechanism.

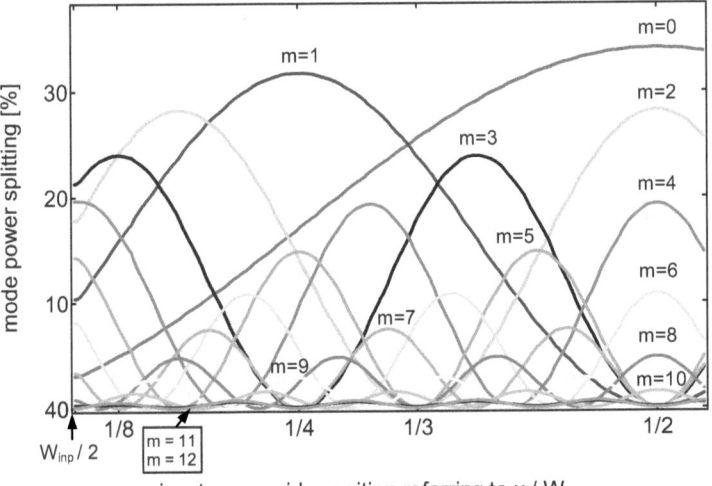

Fig. 4.2 Mode power splitting in percent of the fundamental mode, which is launched into the central MMI coupler section (H = 4 µm, H-h = 2 µm, W_{mmi} = 30 µm). The fundamental mode is provided by a rib waveguide (H = 4 µm, H-h = 2 µm, W_{inp} = 5 µm) and moves from the edge to the MMI coupler centre.

Furthermore, the mode power distribution for different input waveguide positions (inner and outer region) will be clearly different.

The number of guided modes in a multi-mode waveguide is proportional to the effective MMI coupler width. Calculations with the mode solver of FIMMWAVE show that in 4µm SOI rib waveguides an increase in coupler width of 2-3 microns (H = 4.0 µm, H-h = 2.0 µm) increase the number of guided modes by one (see Fig. 4.3 a). This is in good agreement to the calculation of mode numbers by the guiding parameter of a waveguide V [39]. V can be written as $k_0 \cdot W \cdot \sqrt{n_r^2 - n_c^2}$ with the free space wave number $k_0 = 2\pi/\lambda$, the waveguide width W and the effective refractive indices of a planar SOI waveguide (n_r) and surrounding cladding (n_c). The number of guided modes is then given by V/π.

The number of modes becomes relevant, if we compare ideal propagation constants in multi-mode waveguides with real propagation constants. The ideal mode spectrum is given by [38]:

$$\Delta \beta = \beta_0 - \beta_m = \frac{m(m+2)\pi}{3L_\pi} \quad (4.4)$$

This spectrum is obtained under the assumption of $k_{x,m} \approx (m+1)\pi / W_{eff,mmi}$. Here, $k_{x,m}$ is the lateral wave number of mode m and W_{eff} the effective width of the multimode waveguide. Following [38], the approximation is only valid for high contrast waveguides with $W_{eff,mmi} = W_{mmi}$. Therefore, we have to expect phase errors in lateral low index waveguide systems that can lead to blurring of the output images by non-ideal interference of the different modes [40].

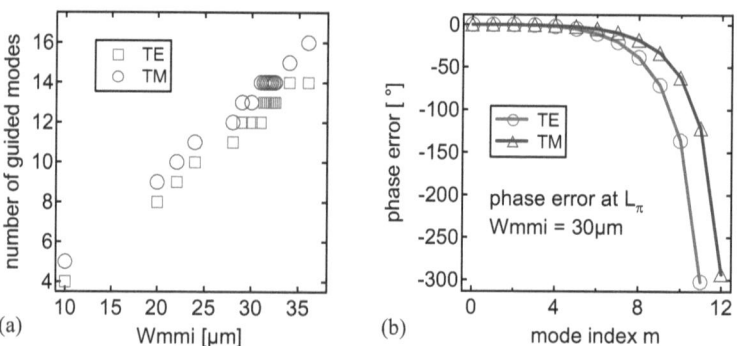

Fig. 4.3 Number of guided modes in the central section as function of MMI coupler width for TE and TM modes (a). The plot in (b) shows phase errors as function of mode index for W_{mmi} = 30µm at L_π.

The phase errors of a MMI coupler in SOI rib waveguide technology (H = 4 µm, H-h = 2 µm, W_{mmi} = 30µm,) at a length of L_π are shown in Fig. 4.3 (b). The errors are calculated by:

$$\Delta\varphi_{error} = (\beta_0 - \beta_m)L_\pi - \frac{m(m+2)\pi}{3} \qquad (4.5)$$

Here, the propagation constants ($\beta_{0...m}$) result from simulations. The calculated phase errors show a deviation from the ideal mode spectrum at higher order modes. Furthermore, the modal phase errors are higher for TE mode than for TM mode.

An option to reduce the effect of phase errors is the reduction of mode power in higher order modes. This can be achieved by a suitable choice of input waveguide width. Fig. 4.4 and Fig. 4.5 show simulated mode power distribution versus the input waveguide width for excitation at $W_{mmi,eff}/3$ (adapted to 2×2 MMI coupler) and at $W_{mmi,eff}/8$ and at 3/8 $W_{mmi,eff}$ (adapted to a 4×4 MMI coupler). The geometrical MMI coupler width W_{mmi} is 30 µm. For excitation at $W_{mmi,eff}/3$, the power in the two lowest order modes increases continuously, if we increase the input waveguide width. The mode power of other higher order modes decreases at larger input waveguide width.

The mode power distributions for an input waveguide position at 1/8 and 3/8 of effective W_{mmi} (Fig. 4.5 a, b) show the same behavior. Finally, the concrete mode power distribution will also depend on the individual excitation of the modes, i.e. on the actual input waveguide position.

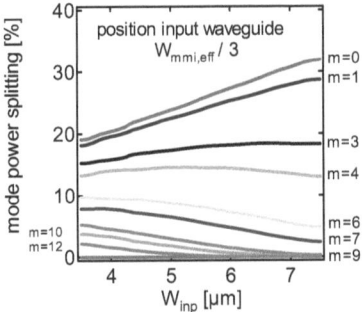

Fig. 4.4 Mode power splitting for restricted interference (W_{mmi} = 30µm) as function of the input waveguide width W_{inp}. The modes at m = 2, 5, 8 and 11 are not excited by fundamental mode of the input waveguide corresponding to paired interference.

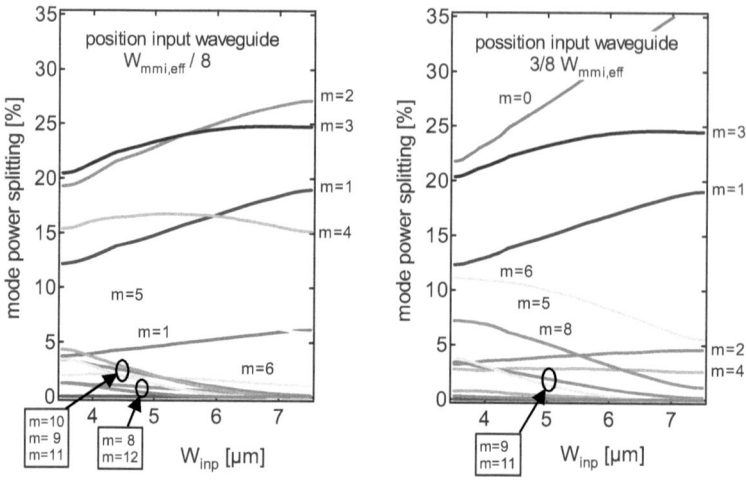

Fig. 4.5 Mode power splitting at access waveguide positions of $W_{mmi,eff}/8$ and $3/8\ W_{mmi,eff}$ as function of the input waveguide width W_{inp}.

A decrease of power in higher order modes can therefore be achieved by widening the input waveguides. However, the increase of all access waveguides width is limited by the possibility of coupling between the waveguides. Furthermore, the gap between the access waveguides decreases by widening of the waveguides. This is only tolerable as long as fabrication tools can resolve the gap between the waveguides. An access waveguide width of ~ 5.5 µm should be sufficient to reduce interference degradation caused by phase errors of higher order modes. This result underlines the importance of access waveguide geometry for design of high performance MMI couplers.

Optimization of excess loss and imbalance

An important characteristic of a MMI coupler is the total guided mode power (P_{tot}) in the central MMI section. A consideration of total guided power should allow for estimation of excess loss and imbalance dependent on coupler geometry. Fig. 4.6 (a,b) show the total mode power of various MMI coupler and input waveguide widths.

The total mode power is obtained by summing up the power of all individual guided modes. Power values are normalized with respect to the input power.

Here, the input rib waveguide width was chosen as 3.5 µm. The dependence of total mode power on MMI coupler width (Fig. 4.6 a) is relatively weak. The complete mode power of guided modes is about 94% independent of coupler width (corresponding to ~ 0.3 dB excess loss). The missing 6% is lost due to radiation modes and reflections. The ratio of the participating loss mechanisms cannot be explicitly identified by the used FIMMPROP tool. However, the low refractive index change between input waveguide and central MMI coupler section indicate a rather low influence of reflections.

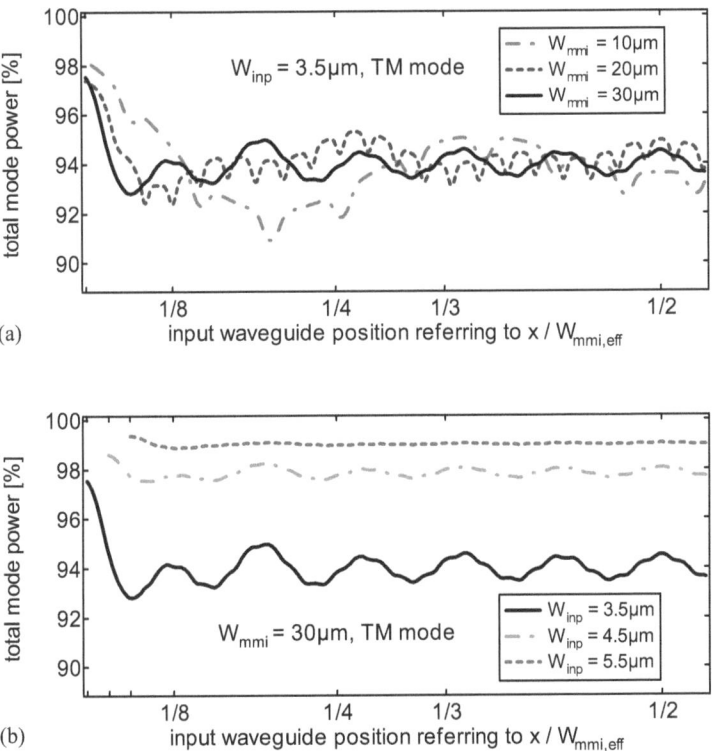

Fig. 4.6 Total mode power (P_{tot}) dependent on effective MMI coupler width (a) and input waveguide width (b). The input waveguide position shifts from the MMI coupler edge to $W_{mmi,eff}$ / 2. TE polarization shows generally slightly less total mode power.

However, at constant MMI coupler width a clear change of total mode power can be observed (e.g. $W_{mmi} = 30$ μm) if the input waveguide width varies (Fig. 4.6 b). For an input waveguide width of 5.5 μm, the total mode power reaches values of about 99% (or ~ 0.05 dB excess loss, which will be twice by use of in- and output waveguide). The available MMI coupler mode power benefits therefore from an expanded input waveguide width. With respect to polarization the simulations show a slightly higher total mode power for TM than for TE mode. For final MMI coupler designs in this thesis, an access waveguide width of 5.5μm was used.

The figures allow also conclusions with regard to coupler imbalance as function of coupler geometry. For this purpose we consider the power difference between total mode power at given input waveguide positions 1 and 2. Fig. 4.7 (a) shows this power difference in total mode power for input waveguide positions at the edge (position 1) and at the centre of the MMI coupler (position 2). The power difference is plotted in decibel after: $10 \log (P_{tot, input\ position1}/P_{tot,\ input\ position2})$. Starting with relatively high difference, the difference decreases clearly for TE and TM polarized light with increasing input waveguide width. For input width of $W_{inp} \geq 5.5$ μm the difference falls below 0.025 dB. If we consider input waveguide position 1 at 1/8 and position 2 at 3/8 of the effective MMI coupler width (Fig. 4.7 b), the difference is already ≤ 0.025 dB for an input waveguide width of ~ 4 μm. Therefore, the position of the outer waveguides should not be chosen at the edge of MMI coupler for the design of a balanced 4×4 devices.

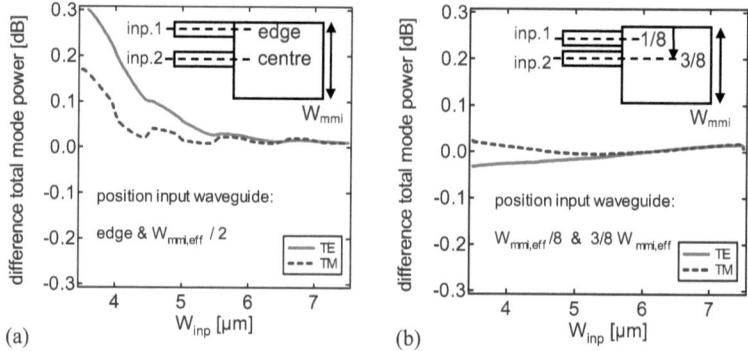

Fig. 4.7 Difference of total mode power (P_{tot}) as function of different input waveguide positions. The input waveguide positions are on the edge / centre of MMI coupler in (a). The input waveguide positions in (b) are at $W_{mmi,eff}/8$ and $3/8\ W_{mmi,eff}$. The MMI coupler width W_{mmi} is 30 μm.

Beat lengths in different SOI technologies

The polarization dependence of MMI couplers is closely related to the SOI waveguide technology. Fig. 4.8 shows different waveguide types in SOI technology. One example is a deep-etched rib waveguide in 4 µm SOI material, which is shown in Fig. 4.8 (b). Here, the MMI coupler is laterally etched to the buried oxide (BOX) plane and able to exceed MMI coupler performance of standard rib-waveguide technology due to the high-index contrast in lateral direction allowing strong confinement of optical modes. This confinement involves reduction of phase errors as shown for 1×4 MMI couplers in [41; 42]. Furthermore, deep etch technology skips technology issues normally given by etch-depth tolerances, which influence the index contrast and therefore the effective waveguide width. First calculations of phase errors in the different SOI waveguides indicate a relation between phase errors and the MMI coupler width. This behaviour is so far not precisely investigated. An increased MMI coupler width tends to result in a reduction of the phase errors. The calculations were carried out by the use of FIMMWAVE mode solver.

However, the polarization dependence of MMI couplers can be described by the polarization dependent beat length L_π, which is defined by $\pi/(\beta_0-\beta_1)$. Here, β_0 and β_1 denote the propagation constants of the fundamental and the first order mode in a SOI waveguide. Tab. 4.2 compares the calculated beat lengths of the different SOI waveguide types. The waveguides in nano-technology (W = 10µm) show the highest polarization dependence, i.e. the highest beat length difference by use of TE and TM polarized light.

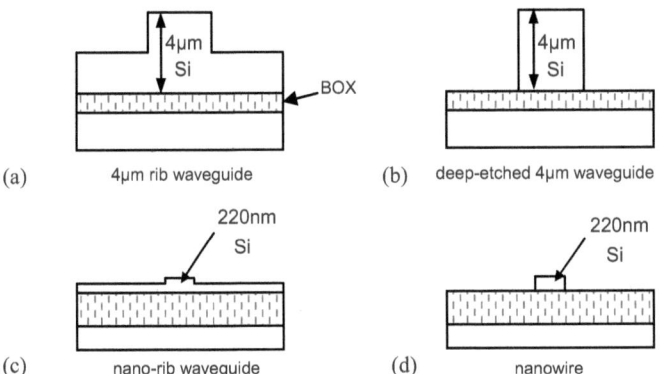

Fig. 4.8 Waveguides in different SOI waveguide technologies: (a) 4µm rib waveguide, (b) nano-rib waveguide, (c) deep-etched 4µm waveguide and (d) nanowire.

Tab. 4.2 Beat lengths ($L_\pi = \pi/(\beta_0-\beta_1)$) for TE and TM polarized light in different waveguide technologies. The beat length calculation in 4µm SOI technology was carried out at a waveguide width of 30 µm. The waveguide width in nano-technology is reduced to 10 µm due to the increased number of guided modes compared to 4µm SOI technology.

SOI Waveguide technology	4µm rib	4µm deep-etched	220nm nano-rib	220nm nanowire
Waveguide geometry	H=4.0µm, W=30µm, H-h=2µm	H=4.0µm, W=30µm, H-h=4µm	H=0.22µm, W=10µm, H-h=0.07µm	H=0.22µm, W=10µm, H-h=0.22µm
beat length TE mode ($L_{\pi,TE}$)	2772.8 µm	2694.3 µm	254.8 µm	243.5 µm
beat length TM mode ($L_{\pi,TM}$)	2785.1 µm	2715.3 µm	172.1µm	182.4 µm
normalized beat length difference ($L_{\pi,TM}/L_{\pi,TE}$ -1)·100%	0.4 %	0.8 %	-32.5 %	- 25.1 %

Here, the beat lengths for TM polarized light are clearly shorter than for TE polarized light. This limits the optimum device performance in nano-technology to one polarization.

Deep etched 4 µm SOI waveguides (W = 30µm) have about the same polarization dependence as 4 µm rib SOI waveguides (W = 30µm). However, the slight increase of the beat length difference in case of deep etched 4µm SOI waveguides may result in device performance degradation in particular for MMI coupler devices with higher complexity as given by 4×4 MMI coupler.

The calculations on 4µm SOI rib waveguides indicate also that the beat length difference decreases with smaller MMI coupler width. For example a MMI coupler width of 20 µm leads to beat lengths of 1253.1 µm and 1260.1 µm for TE and TM polarized light, respectively. These values correspond to a normalized beat length difference of 0.56 %.

4.2 Two selected MMI couplers

The last section provided design guidelines for the MMI couplers. As shown in Fig. 4.6, the total mode power of an MMI coupler in 4µm SOI rib waveguide technology can be controlled by the ratio W_{inp} / W_{mmi}. The coupler width W_{mmi}= 30 µm and the input waveguide W_{inp} = 5.5 µm (W_{inp} / W_{mmi} ~ 0.18) are good starting values for the final design of a 4×4 MMI coupler due to high total power in the central MMI section and the possible arrangement of 4 access (in- and output) waveguides at $W_{mmi,eff}$/8, 3·$W_{mmi,eff}$/8, 5·$W_{mmi,eff}$/8, 7·$W_{mmi,eff}$/8. These waveguide positions result in a gap of ~ 2 µm between the access waveguides, which should be sufficient to avoid impairments due to fabrication limits. The minimum access waveguide separation should not be smaller than 2 µm.

The final 2×2 MMI coupler design was carried out at a smaller MMI coupler width. This results from the requirement of only two access (in- and output) waveguides. A MMI coupler width of 22 µm and an arrangement of the access waveguides at $W_{mmi,eff}$/3, 2·$W_{mmi,eff}$/3 leads to a gap of ~ 2 µm. Here, the ratio W_{inp} / W_{mmi} is about 0.25 and the number guided TE and TM modes 9 and 10, respectively.

The optimization of the 2×2 and 4×4 MMI couplers comprises a fine-tuning of the MMI coupler geometry with regard to low excess loss and imbalance. After "rough" definition of the MMI coupler width and access waveguide width above, the fine-tuning is based on a variation of the input waveguide positions. This includes the record of all output waveguide transmissions at a given input waveguide position as function of MMI coupler length. The analysis of the output data (loss, imbalance) leads to a final determination of the in- and output waveguide position and the MMI coupler length.

In case of the 2×2 MMI coupler, the best performance was obtained at access waveguide positions with separation slightly less than 2 µm. Therefore, the 2×2 MMI coupler width of initially 22 µm was slightly widened to preserve the relative location of the access waveguides to the central MMI coupler section. In the case of the 4×4 MMI coupler, the better performance is formed if the input waveguides are not placed equidistantly (1/8 $W_{mmi,eff}$), but with a slight offset. Therefore, also the 4×4 MMI coupler width was slightly increased. The final widths of 2×2 and 4×4 MMI coupler were chosen with 22.1 µm and 32 µm, respectively. The length dependent excess loss and the excess loss of the 2×2 MMI coupler is shown in Fig. 4.9 (a, b) for TE and TM polarized light.

The minimum excess loss is slightly higher for TM polarized light. The excess loss minimum is located at a coupler length of 780 µm (0.17 dB TE, 0.21 dB TM).

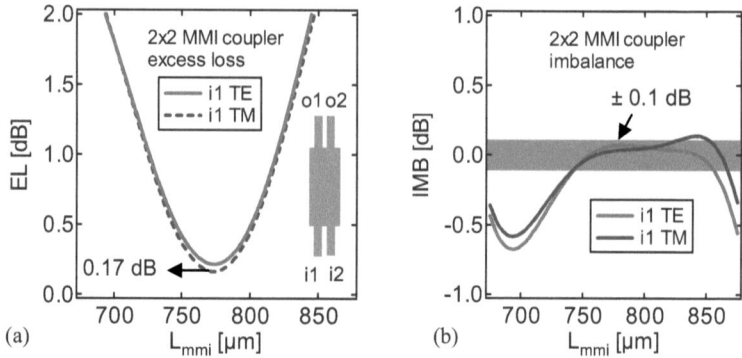

Fig. 4.9 Length dependent excess loss (a) and imbalance (b) of a 2×2 MMI coupler in 4μm SOI technology. The input and output access waveguides are denoted with i1, i2, o1 and o2 (see scheme). Due to symmetry, the results for excess loss and imbalance are equally for input 1 and input 2. Therefore, the results are only presented for launched power at input 1. The coupler width W_{mmi} is 22.1 μm.

A deviation from that length leads rapidly to higher excess loss due to escape of locally maximized MMI coupler images. The imbalance is minimal at a coupler length of 785 μm (~ 0.02 dB). With respect to balanced operation, the length of 785 μm was finally used in 2×2 MZ-DI layouts.

The excess loss of a 4×4 MMI coupler is shown in Fig. 4.10 for TE (a) and TM polarized light (b). The polarization dependent excess loss is generally defined by $10 \cdot \log[(P_{o1}+P_{o2}+P_{o3}+P_{o4})/P_{ref}]$. Here, P_{ref} corresponds to the launched power at MMI coupler input 1 (i1) or input 3 (i3). The excess loss minimum of both polarizations at L_{mmi} = 2400 μm is slightly higher compared to the 2×2 MMI coupler.

Furthermore, the excess loss shows a dependence on the input waveguide index, i.e. excess loss figures differ for input waveguide 1 and 3 (i1 and i3). The excess loss beyond optimum coupler length increases faster for input 3.

The simulated imbalance of the 4×4 MMI couplers is shown in Fig. 4.10 (c, d). The imbalance behavior of the 4×4 MMI coupler is more complex due to the MMI coupler asymmetry. Symmetry exist only for the outer (i1 and i4) and the inner ports (i2 and i3). The MMI coupler length of minimum excess loss is not equal to the optimum length for lowest imbalance, which is larger than 2400 μm. At L_{mmi} = 2425 μm, the imbalance is 0.18 dB for TE and 0.11 dB for TM polarized light.

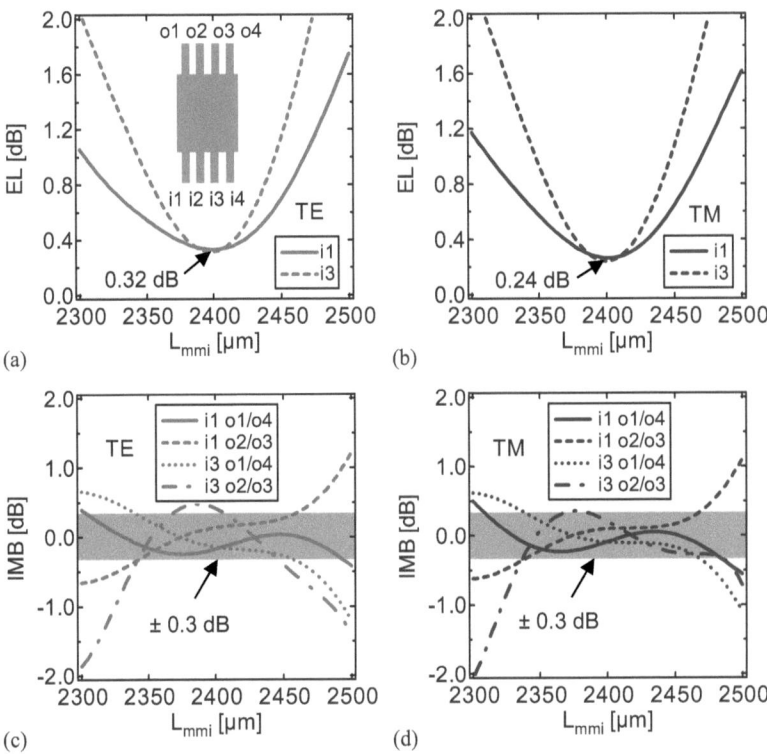

Fig. 4.10 Excess loss (a, b) and imbalance (c, d) of a 4×4 MMI coupler for TE and TM polarized light. The excess loss and imbalance is calculated by use of input 1 (i1) and input 3 (i3). Input 2 (i2) shows the same results as input 3 due to symmetry. Correspondingly, the results by use of input 1 and input 4 are the same. The imbalance plots base on consideration of the π-shifted output ports 1 and 4 (denoted by o1/o4) and output ports 2 and 3 (denoted by o2/o3). The MMI coupler width W_{mmi} during the calculations is 32 µm.

A further characteristic of MMI couplers is their relative output phase. The ideal relative phase shifts (between one input and one output waveguide) can be written in a very compact form with Besse [43]. The ideal relative phases of 2×2 and 4×4 MMI couplers were already presented in terms of transfer matrices in section 2.3. When using 2 input ports, we expect a differential phase shift, which is ideal π in case of a 2×2 MMI coupler.

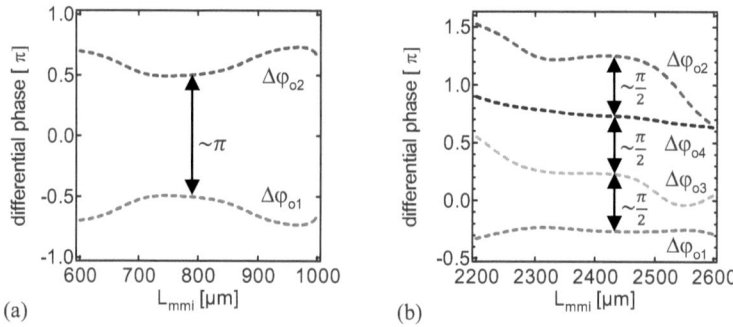

Fig. 4.11 Differential phases of a 2×2 (a) and a 4×4 (b) MMI coupler as a function of coupler length. The phase difference of ~π and ~π/2 are close to the theoretically expected values.

The ideal differential phase shift is defined by $[\Delta\varphi_{o2} - \Delta\varphi_{o1}] = [(\varphi_{i1o1} - \varphi_{i2o1}) - (\varphi_{i1o2} - \varphi_{i2o2})] = \pi$. The differential phases at output 1 and 2 are denoted by $\Delta\varphi_{o1}$ and $\Delta\varphi_{o2}$, respectively. For the 4×4 MMI coupler, we obtain with [43] ideal differential phase shifts are π/2. The differential phases are defined similarly to the 2×2 MMI coupler ($\Delta\varphi_{o1} = \varphi_{i1o1} - \varphi_{i3o1}$, $\Delta\varphi_{o2} = \varphi_{i1o2} - \varphi_{i3o2}$, $\Delta\varphi_{o3} = \varphi_{i1o3} - \varphi_{i3o3}$, $\Delta\varphi_{o4} = \varphi_{i1o4} - \varphi_{i3o4}$). Fig. 4.11 shows calculated differential phases at the 2×2 and 4×4 MMI output ports for TE polarized light. The results for TM polarized light show the same characteristic.

We obtain the predicted phase characteristics for both MMI couplers close to the coupler length for optimum excess loss and imbalance. The 2×2 MMI coupler provides in differential operation a π- shift between the outputs. The differential phase shifts of a 4×4 MMI coupler depend on the compared output waveguides (Fig. 4.11 b). Here, input waveguides 1 and 3 are used.

MMI coupler bandwidth

The achievable bandwidth of 2×2 and 4×4 MMI couplers was determined by wavelength dependent calculations. A first study investigated excess loss as a function of wavelength. The results are shown in Fig. 4.12. The 2×2 MMI coupler geometry during the calculation is as follows: H = 4.0 μm, H-h = 2.0 μm, W_{mmi} = 22.1 μm, W_{inp} = 5.5 μm, L_{mmi} = 780 μm. The 4×4 MMI coupler has the following dimensions: H = 4.0 μm, H-h = 2.0 μm, W_{mmi} = 32 μm, W_{inp} = 5.5 μm, L_{mmi} = 2425 μm. The excess loss of the 2×2 MMI coupler never exceeds 0.3 dB over the C-band (1530 - 1565 nm), although the loss increases at the boarders of the band. Over a wavelength range of 200 nm (1450 nm - 1650nm) the loss increases up to ~ 1 dB.

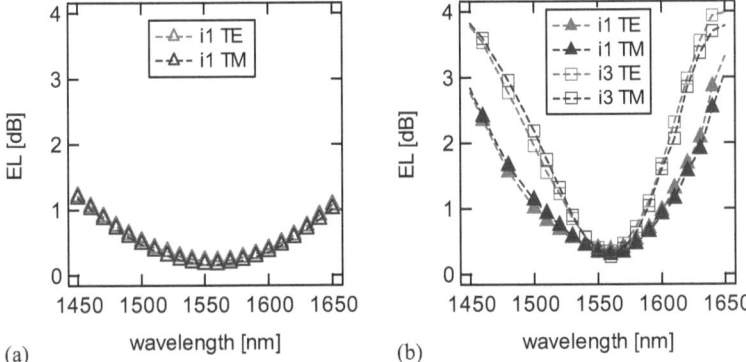

Fig. 4.12 Excess loss as function of wavelength for a 2×2 (a) and 4×4 MMI coupler (b) in 4µm SOI rib waveguide technology. The excess loss of the 4×4 MMI coupler for light at input 3 (i3) is higher compared to light at input 1 (i1).

The 4×4 MMI coupler shows the same behavior at the boarders of the C-band, but with higher losses in particular for input 3. In this case, the loss is approximately 0.9 dB. Besse [44] predicted a 1/N dependence of the available bandwidth of MMI couplers. This is confirmed for input 1. However, if we consider input 3, the available bandwidth is further reduced by a factor of 2. Another calculation has been carried out to evaluate the wavelength dependence of the coupler imbalances.

The calculated imbalances of the 2×2 and 4×4 MMI couplers for wavelengths of 1530 nm - 1580 nm are shown in Fig. 4.13. It is obvious that the imbalance values of the 4×4 MMI coupler exceed the imbalance values of the 2×2 MMI coupler. Similar to excess loss, input 3 (i3) of the 4×4 MMI coupler suffers from slightly higher imbalance. As will be shown later, the MMI coupler imbalance of ≤ 0.5 dB leads to MZI extinction ratios better than 20 dB. The simulation therefore indicates proper C-band performance of 4×4 MMI couplers. TE- and TM-polarized light show the same characteristics.

Additionally, the relative phases at the MMI coupler outputs shall be considered. Fig. 4.14 shows plots of the differential phase shifts of a 2×2 and a 4×4 MMI coupler. The differential phase shifts are approximately constant over a wavelength range of 50 nm, i.e. $[\Delta\varphi_{o1} - \Delta\varphi_{o2}] \sim \pi$ for the 2×2 MMI coupler and $[\Delta\varphi_{o3} - \Delta\varphi_{o1} \sim \pi/2, \Delta\varphi_{o4} - \Delta\varphi_{o1} \sim \pi, \Delta\varphi_{o2} - \Delta\varphi_{o1} \sim 3\pi/2]$ for the 4×4 MMI coupler. The calculated phase accuracy is better than 5° with respect to the ideal differential phase shift over C-band.

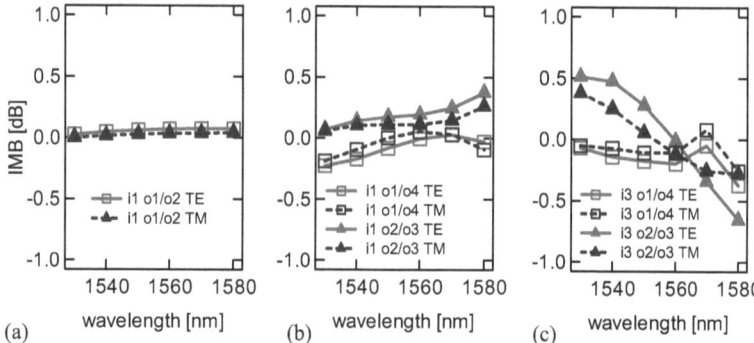

Fig. 4.13 Imbalance of a 2×2 (a) and a 4×4 MMI couplers (b,c) over wavelength range of 1530 nm – 1580 nm. The input power is launched at input 1 (i1) in case of the 2×2 MMI coupler and at input 1 (i1) and input 3 (i3) in case of the 4×4 MMI coupler.

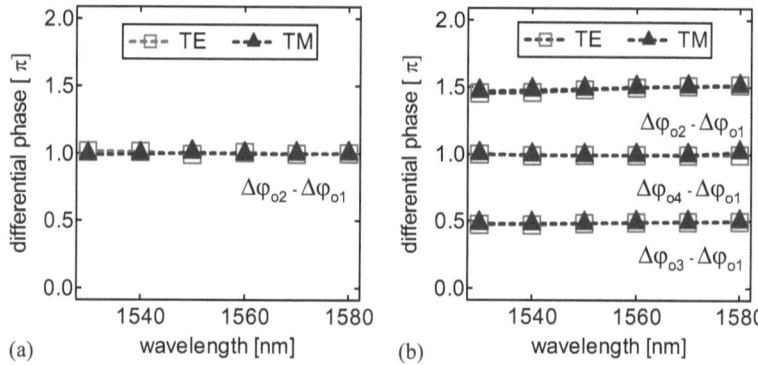

Fig. 4.14 Differential phase shifts at the outputs of a 2×2 and a 4×4 MMI couplers over wavelength range of 1530 nm – 1580nm. The differential phase shifts at the outputs are calculated with respect to differential phase at output port 1 (o1).

Tab. 4.3 summarize the simulated characteristics of the 2×2 and the 4×4 MMI coupler. Also included are results of 1×2 and 1×4 MMI couplers, which are characterized in the same way. The design of 1×N MMI coupler benefits from symmetrical output power distribution. Therefore output port imbalance is intrinsically low. Nevertheless, the imbalance of the 2×2 MMI coupler can be better than 0.1 dB over S-, L- and C-band.

Excess loss and imbalance of the 4×4 MMI coupler do not exceed 0.9 dB and 0.5 dB, respectively. We conclude, SOI MMI couplers show suitable characteristics for applications in the telecom wavelength range.

Tab. 4.3 Summarized calculation results of MMI couplers in 4µm SOI rib waveguide technology. The item "phase accuracy" refers to the differential phase shifts between the MMI coupler outputs compared to nominal differential phase shifts (multiples of $\pi/2$). The symbol W_{inp} denotes the rib waveguide width of all in- and output waveguides.

Coupler / Characteristic	2×2 MMI	4×4 MMI	1×2 MMI	1×4 MMI
Device geometry	H = 4.0 µm H-h = 2.0 µm L_{mmi} = 780 µm W_{mmi} = 22.1 µm W_{inp} = 5.5 µm	H = 4.0 µm H-h = 2.0 µm L_{mmi} = 2425 µm W_{mmi} = 32 µm W_{inp} = 5.5 µm	H = 4.0 µm H-h = 2.0 µm L_{mmi} = 582 µm W_{mmi} = 22.1 µm W_{inp} = 5.5 µm	H = 4.0 µm H-h = 2.0 µm L_{mmi} = 595 µm W_{mmi} = 32 µm W_{inp} = 5.5 µm
Excess Loss C-band	< 0.3 dB	< 0.9 dB	< 0.2 dB	< 0.2 dB
Imbalance C-band	< 0.1 dB	< 0.5 dB	< 0.01 dB	< 0.01 dB
Polarization dep. loss C-band	< 0.1 dB	< 0.2 dB	< 0.01 dB	< 0.02 dB
Phase accuracy better than 5° C-band	yes	yes	yes	yes

4.3 Experimental results

The testing of fabricated 2×2 and 4×4 MMI couplers was carried out with the measurement setup as presented in section 3.5. The 2×2 and 4×4 MMI coupler geometries correspond to the optimized values as presented in Tab. 4.3. Transmission measurements of 2×2 MMI coupler and 4×4 MMI couplers are shown in Fig. 4.15 (a-c) for TE polarized light. With TM polarization, essentially the same results were obtained.

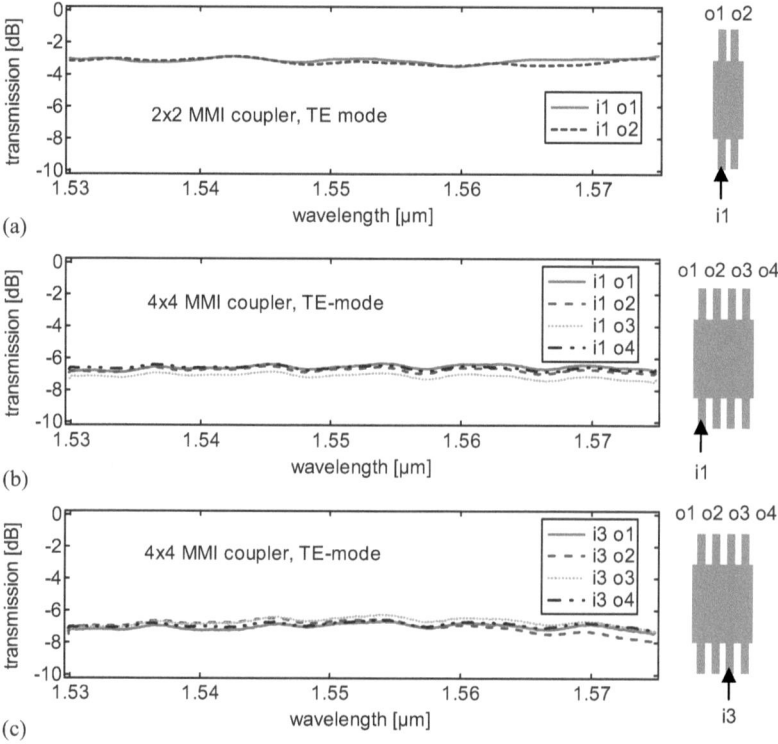

Fig. 4.15 Transmission measurements of 2×2 (a) and 4×4 (b,c) MMI couplers in 4µm SOI rib waveguide technology for TE polarized light. The 2×2 MMI coupler transmissions by use of input 2 (i2) are equal to the transmissions following by use of input 1(i1) due to coupler symmetry. Therefore, the results for input 2 are not shown here. The 4×4 MMI coupler transmissions are shown for two inputs (i1, i3). The results for i4 and i2 correspond to the results by use of i1 and i3, respectively.

The transmission measurements refer to the measurement of a straight waveguide, which is available on each chip containing MMI couplers. Therefore, the measurements show only additional losses arising from MMI coupler excess loss and increased waveguide length. However, the increase in waveguide length will be negligible due to the low-loss characteristic of 4µm SOI rib waveguides by optimum process conditions.

The 2×2 and 4×4 MMI couplers show homogenous transmission over C-band with the expected power splitting of 3 dB and 6 dB, respectively. The imbalance of the 4×4 MMI coupler is slightly higher than for the 2×2 MMI coupler. The excess loss of 2×2 and 4×4 MMI couplers was separately determined by cascading of MMI couplers as schematically shown in Fig. 4.16. A maximum number of 16 2×2 MMI couplers and 7 4×4 MMI couplers was used.

Fig. 4.17 shows the result. The plotted loss values are averaged over C-band. Additionally, linearly fitting curves are included. The measured 2×2 and 4×4 MMI couplers show less polarization dependence. According to calculations, the excess loss of the 2×2 MMI coupler is less than for the 4×4 MMI coupler. The use of input 3 of the 4×4 MMI coupler leads to about 0.23 dB higher excess loss compared to input 3. Tab. 4.4 collects the experimental performances of the considered MMI coupler types, additionally for 1×2 and 1×4 MMI coupler, which are not plotted here. Imbalance and PDL of the transmission measurements were determined by measurements as shown in Fig. 4.15.

Calculations indicate better performance of 2×2 MMI couplers compared to 4×4 MMI couplers in terms of excess loss (EL), imbalance (IMB) and polarization dependent loss (PDL). The 1×2 MMI coupler delivers best performance considering imbalance and PDL. Finally, all MMI coupler types show state-of-the-art characteristics.

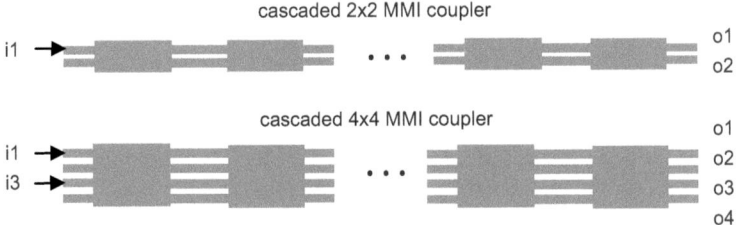

Fig. 4.16 Scheme of the cascaded 2×2 and 4×4 MMI couplers. The waveguides between the MMI couplers have the same length.

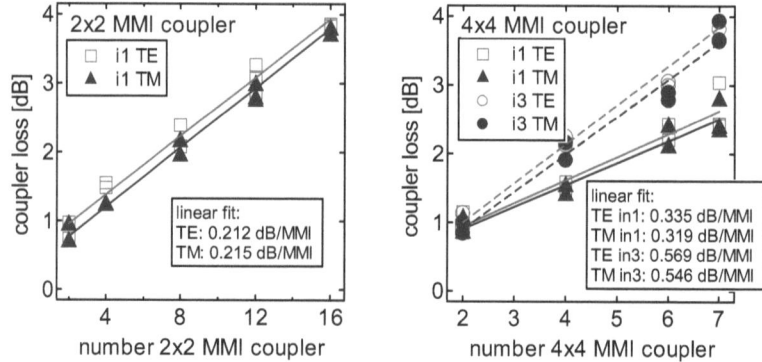

Fig. 4.17 Experimental determined excess loss of cascaded 2×2 (a) and 4×4 (b) MMI coupler for TE and TM polarized light and use of input 1 (in1) and input 3 (in3). Linear fitting curves are used for determination of single coupler loss (dB/coupler). The results are shown in the inserted boxes.

Tab. 4.4 Experimental C-band performance for TE and TM polarized light of different MMI couplers in 4µm SOI rib waveguide technology.

Device / Characteristic	4x4 MMI i1	4x4 MMI i3	2x2 MMI i1	1x2 MMI	1x4 MMI
Excess loss	< 0.4 dB	< 0.6 dB	< 0.3 dB	< 0.3 dB	< 0.4 dB
Imbalance	< 0.7 dB	< 0.7 dB	< 0.2 dB	< 0.1 dB	< 0.3 dB
Polarization dependent loss	< 0.4 dB	< 0.5 dB	< 0.2 dB	< 0.1 dB	< 0.2 dB

The 4×4 MMI coupler shall be compared in more detail to other realizations of 4×4 couplers or 90° optical hybrid devices. There are approaches in III/V, in silica and in LiNbO$_3$. An approach in SOI material with larger cross section [45] (~10.5µm) is presented. The 4×4 MMI device in 4µm SOI technology match the characteristics of previously realized devices in other planar technologies. The 4µm-SOI device shows a clearly better performance compared to the device in 10.5 µm SOI. High bandwidth capability (Fig. 4.15) is an additional argument in favour of the 4µm-SOI approach. The PDL is not mentioned in the published material.

Tab. 4.5 Realized 4×4 or 90° optical hybrids in SOI material and other material systems. Wavelength range and PDL were not always specified.

Material system / Characteristic	III/V Pennings [46] MMI coupler	Silica Doerr [19] Star coupler	LiNbO$_3$ Heidrich [47] Directional coupler network	SOI Wei [45] MMI coupler
Excess loss	≤ 1 dB	≤ 2 dB	≤ 4.2 dB	≤ 3.7 dB
Imbalance	≤ 0.3 dB	"high"	≤ 0.2 dB	≤ 2.3 dB

4.4 Fabrication tolerances of MMI couplers

Realised rib waveguides may show deviations from ideal geometry. Such deviations arise mainly from process variations (lithography, rib etching) and non-uniformities of the top-silicon substrate thickness H. The variations and the non-uniformities result in undesirable degradation of MMI coupler performance and change of geometrical birefringence. Fig. 4.18 illustrates the effects on rib waveguide geometry.

Non-ideal lithography leads to deviations and non-uniformity of the rib waveguide width W across the wafer.

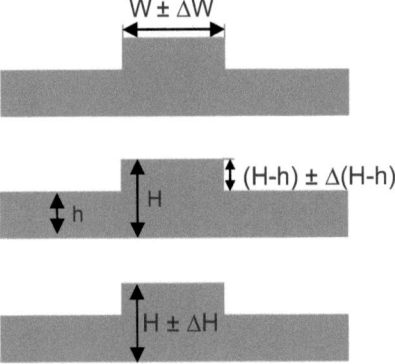

Fig. 4.18 Process variations related to waveguide geometry, i.e. variations in rib waveguide width ΔW, etch depth Δ (H-h) and substrate thickness ΔH.

The non-uniformity is primarily a result of the varying exposure dose of contact lithography across the wafer. Therefore, a mask layout is not exactly reproduced as resist pattern on SOI-wafer. The width variation was studied under use of scanning electron microscope (SEM) measurements. An offset ΔW up to ± 0.2 µm of the nominal rib waveguide width $W = 3.5$ µm across the wafer was observed. Fig. 4.19 (a) shows exemplary the width variation of a single Omega-MZI device. The rib waveguide width varies in the range of about 0.2 µm.

A nominal rib etch depth (H-h) of 2 µm can vary by approximately + 0.3 µm (+ 15 %) towards the edge of the wafer, as measurements with stylus profiler (Dektak) have been shown. An exemplary measurement is plotted in Fig. 4.19 (b).

The thickness of the top silicon H and the buried oxide (BOX) should be 4 µm and 1 µm, respectively. However, manufacturers of BESOI wafers specify the top silicon thickness with $H = 4 \pm 0.5$ µm (± 12.5 %).

A fabrication tolerance analysis of 2×2 MMI couplers was carried out with the FIMMPROP-tool of Photon Design. Besides the impact of variations in MMI coupler width, etch depth and substrate thickness, also variation in length shall be considered here. An estimation of length related effects can be made by reference to Fig. 4.20.

A length variation of ± 10µm ($\sim 1\%$) of the nominal $L_{mmi} = 780$ µm leads to an additional excess loss of ~ 0.04 dB (TE and TM mode). The imbalance increases by about 0.02dB. However, a length deviation of 10µm or more due to fabrication tolerances may not be expected. Therefore, the impact of length variation on coupler excess loss and imbalance is rather low.

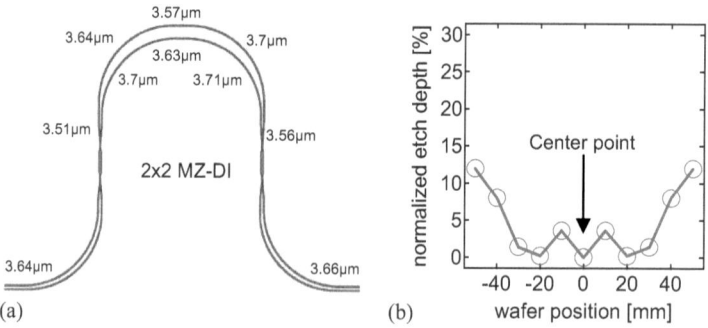

Fig. 4.19 (a) Width variation of rib waveguides in a MZ-DI. (b) Etch depth variation across a 4 inch wafer measured with Dektak.

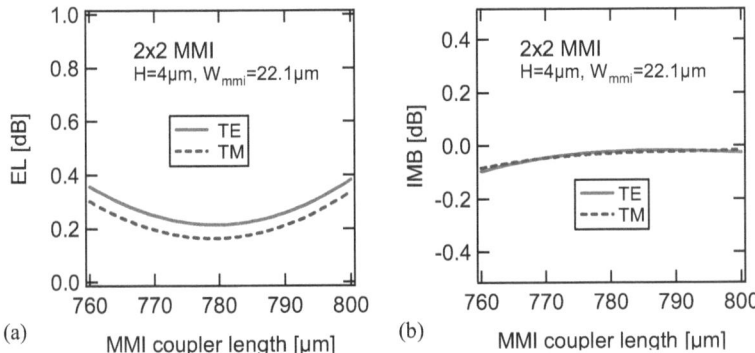

Fig. 4.20 Excess loss (EL) and imbalance (IMB) vs 2×2 MMI coupler length. The nominal coupler is 780µm.

Basically, the quadratic relation between coupler length and width will lead to higher impact in case of coupler width variation [44]. Simulation results concerning width variation ΔW are shown in Fig. 4.21 (a,b). Despite a slight shift, the excess loss curve of TM mode corresponds approximately to the excess loss for TE mode. The width variation ΔW of $\pm 1\%$ (~ 0.2 µm) affects an additional excess loss and imbalance smaller than 0.1 dB, which is acceptable.

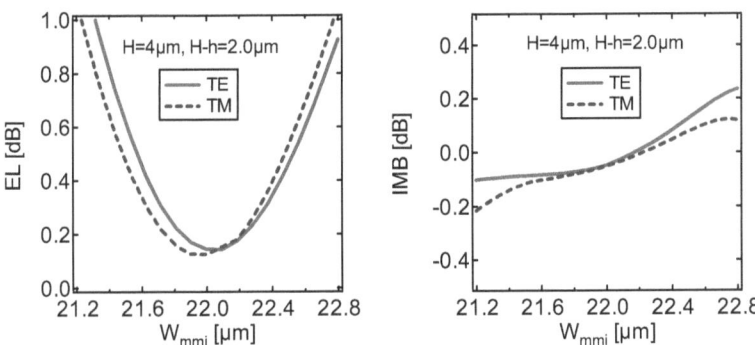

Fig. 4.21 Excess loss (a) and imbalance (b) vs MMI coupler width. The nominal MMI coupler width is 22.1 µm.

However, larger ΔW cause through the parabolic shaped excess loss dependence quickly higher losses. Further simulations were conducted to figure out influence of variation in etch depth. Resulting excess loss and imbalance around the nominal etch depth of 2 μm (on 4 μm SOI substrate) are depicted in Fig. 4.22.

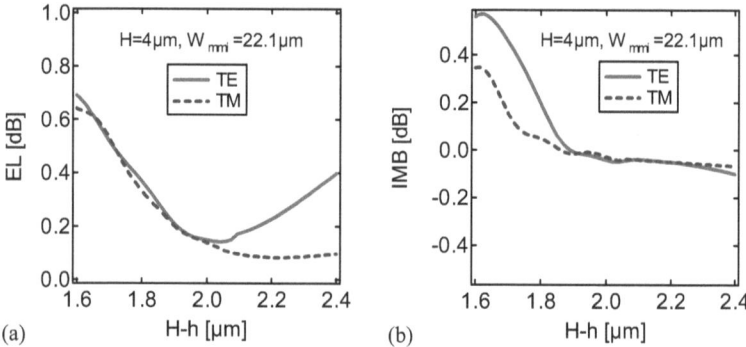

Fig. 4.22 Excess loss (a) and imbalance (b) vs. MMI coupler etch depth variation. The nominal MMI coupler etch depth is 2.0 μm.

At higher etch depth (> 2 μm) the excess loss differs for the two fundamental polarizations. For 10 % (or 0.3 μm) deeper etch the excess loss stays below 0.35 dB for TE mode. In case of TM mode the excess loss remains by about 0.1 dB with increasing etch depth. The imbalance for both polarizations increases only insignificantly. On the other hand, a conservative etch depth of < 2 μm leads to stronger effects. So results a Δ(H-h) of - 0.3 μm in an excess loss of about 0.5 dB. The imbalance increases for TE mode to more than 0.4 dB.

The influence of a variation in substrate thickness is also considered. Fig. 4.23 (a, b) show related simulation results. At an etch depth of 2 μm and shrinking substrate thickness, the excess loss for TE mode increases gradually (e.g. to ~0.27 dB at ΔH = 0.4 μm) and decreases slightly for TM mode. The imbalances change is not significant. As shown in the same figures, the higher excess loss for TE mode can be compensated with a shallower etch step. An etch depth of 1.7 μm at H around 3.6 μm leads to the calculated values for excess loss and imbalance at nominal dimensions.

A substrate thickness of H > 4 μm and an H-h ≤ 2 leads to significantly increased excess loss and imbalance. Finally, all considered variations show degrading effects on MMI coupler performance.

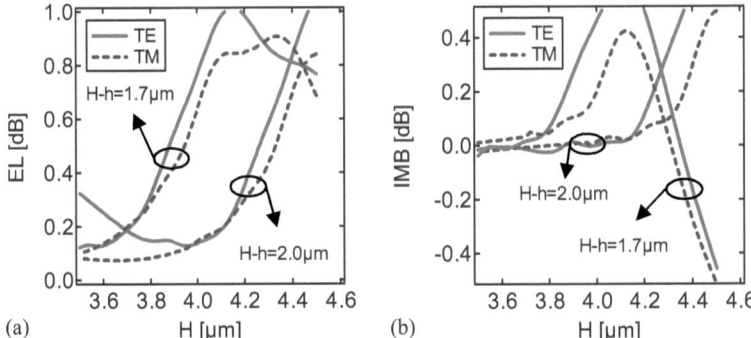

Fig. 4.23 Impact of substrate thickness H variation on 2×2 MMI coupler performance, i.e. excess loss (a) and imbalance (b). The nominal substrate thickness H is 4.0 µm.

From the studied variations of the fabrication process ($\Delta W_{mmi} = \pm 0.2$ µm, $\Delta(H-h) = +0.3$ µm, $\Delta H = \pm 0.5$ µm) the positive substrate variation ($\Delta H > 0$) leads to the biggest deterioration on the MMI coupler performance. Together with a shallow etch serious degradations will occur.

5 Experimental results of Mach-Zehnder delay interferometers

5.1 Layout

The realization of the rib waveguides requires a careful layout under consideration of a high device density and other requirements, i.e. the size of the SOI chip related to limitations in post processing or in the measurement setup. Other requirements can be given by targeted hybrid integration of active components (space requirements, additional layers) or packages with predefined dimensions.

For the layout of the rib waveguide structures, the CAD-tool of RSoft Design Group, Inc. was deployed. The layout of straight waveguides with s-bended section can be carried out in a straightforward way. The waveguide length and width as well as the s-bended section can be defined globally in the RSoft CAD-tool. Fig. 5.1 shows the layout of a 2×2 MMI coupler The MMI coupler layout includes two in- and output waveguides with a typical separation of 50 µm and the MMI coupler with taper and s-bend section.

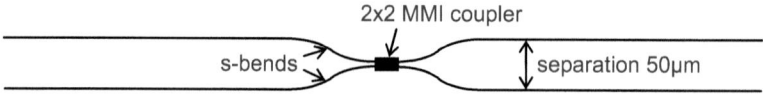

Fig. 5.1 Layout of a MMI coupler in 4µm SOI technology.

A Mach-Zehnder delay interferometer structure can be designed in various ways. This is related to the orientation of the two MMI couplers or the shape of the two interferometer arms (one with delay and phase shifter). Obviously, a larger delay length ΔL implies a larger device size.

Apart from the MMI coupler size and delay length, a determining factor for device size is the minimum radius of bends used for networking. As shown by calculations and experiments (see section 3.6) suitable bending radii of the 4µm SOI rib waveguides are in the range of 5 mm. Along with that we achieve the minimization of polarization dependent waveguide loss, which is also less at this bending radius. We also need to keep in mind that the orientation of the delay interferometer arms may influence stress induced birefringence and related PDFS.

With respect to these constraints, the following constellations were designed and tested:

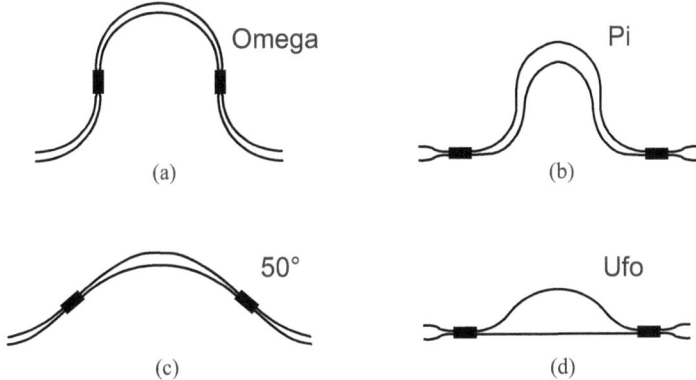

Fig. 5.2 Design of 2×2 Mach-Zehnder delay interferometer in 4μm SOI technology. The inserted alias denotes the names that will be used throughout this thesis.

The individual MZ-DIs shall be shortly described. The Omega- and Pi-type enable the delay implementation over vertical extensions of the outer interferometer arm. They differ in the arrangement of the MMI coupler, which is vertical for the Omega MZI and horizontal for the Pi MZI. Unfortunately, the designs consist of a number of 90° bends which impede the design of space-saving devices.

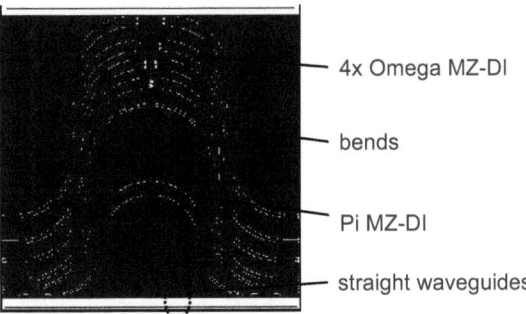

Fig. 5.3 Single-chip layout with MZ-DIs of Omega- and Pi-type. The chip contains also test-structures for MZ-DI independent waveguide characterization.

A clearly smaller footprint is enabled by the 50° and the Ufo design. Here, the two tilted MMIs (50°) and a single bended delay interferometer arm (Ufo) lead to a small MZ-DI device size. Except for the Ufo design, the layout variants allow for the interleaving of MZIs. Fig. 5.3 shows as example a chip layout with MZ-DIs of Pi- and Omega-type. The chip size is 25 mm × 25 mm. On top and bottom are test structures arranged to permit characterization of waveguide loss. The chip coupling to single-mode fibers takes place at the facets on left and right edge of the chip.

5.2   2×2 MZ-DI characterization

A 2×2 Mach-Zehnder delay interferometer consists of two 2×2 MMI couplers. Different MZ-DI designs (Omega, Pi, 50°, Ufo - see Fig. 5.2) were realized and tested. The fabricated chips were separated, polished and coated with an ARC. Wavelength dependent sweeps over the C-band were carried out with an increment of 1 pm. Most of the measurements were carried out with TE- and TM polarized light. For measurement of polarization dependent effects on the devices randomly polarized light was also used.

Extinction ratios, Imbalance and Polarization dependent loss

Fig. 5.4 (a) shows transmission measurements of a 2×2 MZ-DI around 1550 nm with a free spectral range (FSR) of 40 GHz corresponding to $\Delta\lambda_{FSR} \sim 320$ pm for the TE-mode. The in- and output ports of the MZ-DI are denoted by i1, i2, o1 and o2, respectively. The measurements were carried out by use of input 1 (i1) if not differently labelled; see Fig. 5.4 (a). The same measurement is plotted over the C-band (Fig. 5.4 (b).

The graph reveals a uniform loss (~ 1.8 dB) as well as uniform extinction ratio across the entire C-band (minimum extinction 31 dB). TM-polarized light leads to similar characteristics and is therefore not presented here.

The extinction ratios are furthermore analyzed at a chip temperature of 70°C. As can be seen in Fig. 5.5 (a) the extinction ratios remain higher than 30 dB. Therefore, the tuning of the MZ-DI by heating the device will not degrade device performance. In Fig. 5.5 (b) extinction ratios of 90 consecutive measurements of random polarized light are plotted. During the measurement, the Poincaré sphere is scanned by the polarization controller. The extinction ratios spread over a corridor of 7 dB, with minimum extinction ratio of 24 dB.

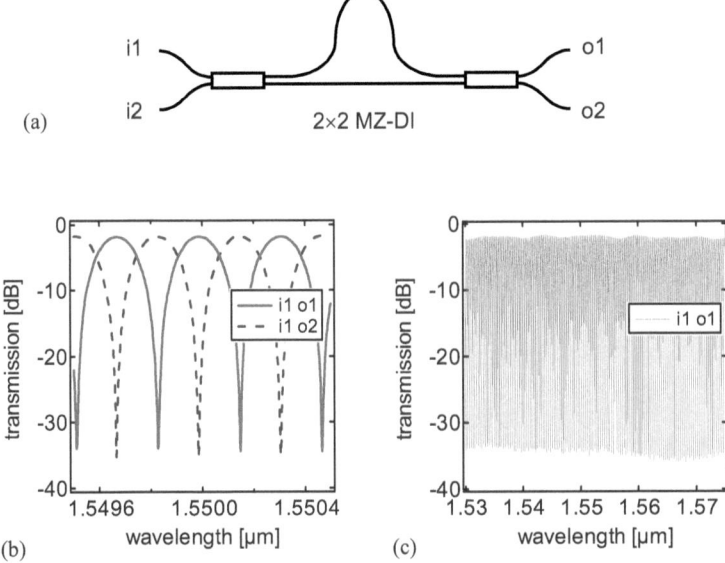

Fig. 5.4 Labeling of the 2×2 Mach-Zehnder delay interferometer used in the experiments (a). Transmissions of a 2×2 Mach-Zehnder delay interferometer around 1550 nm (b) and over C-band (c) for TE polarized light.

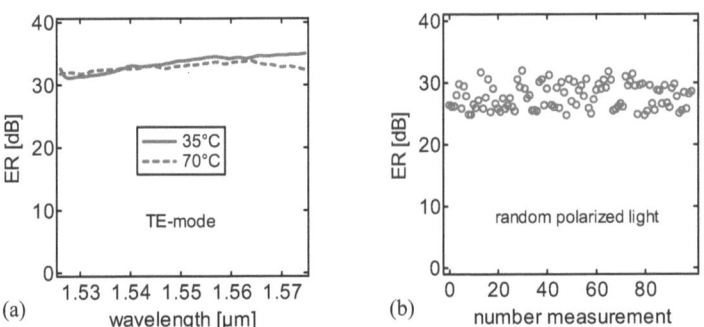

Fig. 5.5 Extinction ratios at single output (o1) of a MZ-DI. (a) shows the extinction ratios over C-band for two temperatures (35°C, 70°C). In (b) the extinction ratios were measured under use of random polarized light (90 measurements).

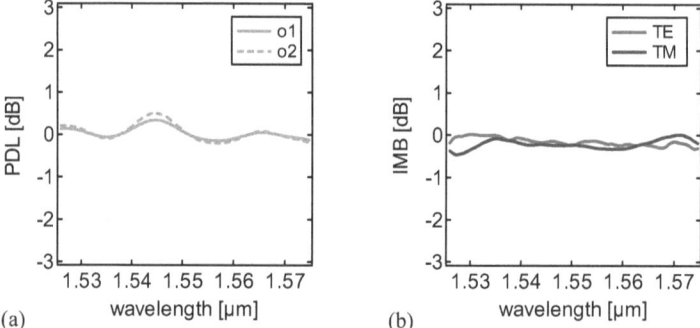

Fig. 5.6 Polarization dependent loss (a) and imbalance (b) of a 2×2 Mach-Zehnder delay interferometer.

Also polarization dependent loss (PDL) and imbalance (IMB) were determined. The PDL is separately defined for each MZ-DI output as a difference of maximum MZ-DI transmission signal by use of TE- and TM polarized light, i.e. (max. $P_{o1,TE}$- max. $P_{o1,TM}$) and (max. $P_{o2,TE}$- max. $P_{o2,TM}$). The imbalance is defined by the signal difference between the output ports 1 and 2 by use of TE- or TM polarized light. Both stay below 0.5 dB across the C-band. This situation is shown in Fig. 5.6 (a, b).

The slightly increased coupler loss and imbalance compared to simulated values of 2×2 MMI-couplers can be attributed to imperfections within the MZ-DI. The MMI couplers are widely apart (~ cm) on the chip surface, which may expose each coupler to different fabrication conditions. A high definition lithography and a more uniform SOI wafer material could improve the coupler characteristics, but would also increase fabrication costs.

Phase accuracy

The accuracy of phase dependent MZ-DI switching can be analysed by means of wavelength dependent transmission measurements. It is convenient to consider the transmission minima of the different ports as reference. Fig. 5.7 (a) shows the results for a 2×2 MZ-DI. The relative phase is calculated with respect to the wavelength at minimum transmission at output 1 by: $2\pi \cdot (\lambda_{min.Po2} - \lambda_{min.Po1})/\Delta\lambda_{FSR}$. Here denote $\lambda_{min.Po1}$ and $\lambda_{min.Po2}$ the wavelength of the transmission minima at output port 1 and 2, respectively. The expression $\Delta\lambda_{FSR}$ corresponds to the wavelength difference of a free-spectral range.

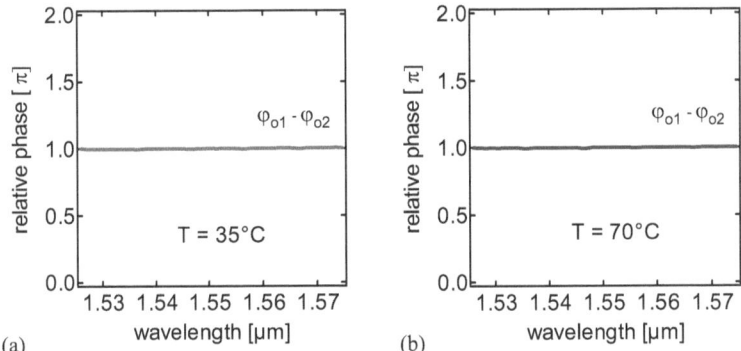

Fig. 5.7 Relative phases at the 2×2 MZ-DI output ports at 35°C (a) and 70°C (b). Phase reference for calculation is output 1.

Of particular interest is furthermore a stable and temperature independent switching behaviour. Thus, the measurements were carried out at two temperatures (35°C, 70°C).

In Fig. 5.7 (a) the phase of the 2×2 MZ-DI remains stable over the C-band with an accuracy of ± 2°. The phases are almost unchanged at temperature of 70°C. This is shown in Fig. 5.7 (b).

Regarding phase accuracy and thermal stability, phase demodulation using MZ-DIs in 4µm-SOI rib waveguide technology is therefore realistic. PDFS measurements were carried out on chips with different 2×2 MZ-DI designs (Omega, Pi, Ufo, 50°). The FSR in all MZ-DI designs is 320 pm. For birefringence tuning, a cladding was applied (~ 200 nm silicon nitride). Finally, the PDFS as shown in Fig. 5.8 was obtained.

As can be seen from Fig. 5.8 the PDFS of the different designs remain within the 1 GHz specifications for DPSK demodulation [14]. Therefore, a very flexible use of the proposed designs will be possible. This is supported by the achieved uniform characteristics with respect to ER, IMB, and PDL.

Another issue is the temperature dependence of demodulator devices. Fig. 5.9 (a) shows PDFS measurements at temperatures of 35°C and 70°C. Here a 2×2 MZ-DI in Omega design was used. The curves show an increase in PDFS, which is smaller than expected. So far the reason for the low temperature dependence of PDFS performance is not fully understood. The results may arise from a low birefringence dependence on temperature, i.e. $dn_{eff}^{TE}/dT \approx dn_{eff}^{TM}/dT$.

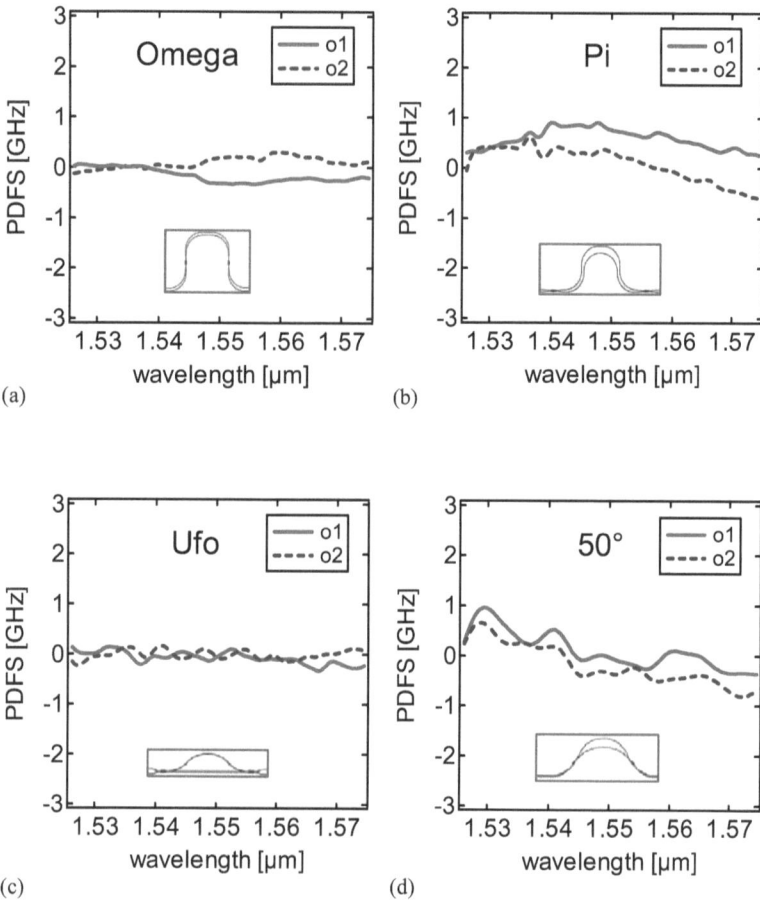

Fig. 5.8 PDFS in GHz over C-band of different 2×2 MZ-DI designs, i.e. a) Omega, b) Pi, c) Ufo, d) 50°. Each PDFS remains in the area of ±1 GHz (±8 pm).

Furthermore, a non-linear relation between temperature and stress-induced effective birefringence is possible. In this case, the generated stress by the cladding might be unchanged as far as the considered temperature range is less than 35 K.

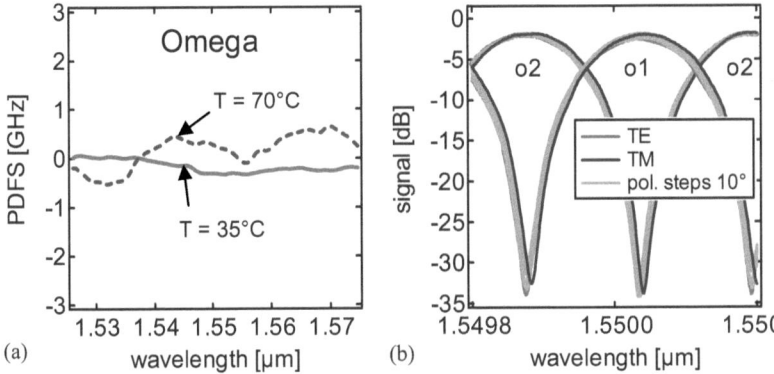

Fig. 5.9 PDFS at 2×2 MZ-DI output 1 (o1) for two different temperatures (35°C, 70°C) covering C-band (a). Transmissions in linear polarization steps of 10° between TE and TM light (b).

PDFS at extended polarizations

PDFS was also measured for other than the two fundamental polarizations, which is the standard definition of PDFS. The experiment varied the angle of the state of polarization. Transmissions of 2×2 MZIs were recorded by steps of 10° between TE and TM polarized light. The results are shown in Fig. 5.9 (b). If we do not refer to TE & TM polarization, extended polarization PDFS can be defined as the maximum shift between the minima of polarization dependent MZI filter curves. PDFS as determined this way is slightly increased compared to the standard definition. The difference is usually not very large. For example, if the standard PDFS is 1 GHz (± 0.5 GHz) we obtain about ± 0.61 GHz in case of the extended polarization PDFS.

5.3 2×4 MZ-DI characterization

Extinction ratios, imbalance and polarization dependent loss

In the 2×4 MZ-DI constellation, one MMI coupler of the 2×2 MZ-DI is replaced by a 4×4 MMI coupler (L_{mmi}=2430 µm) using the MMI coupler inputs 1 and 3. The corresponding MZ-DI matrix calculation shows transmissions at the output ports, which are shifted by 1/4 FSR (or 90°). The labelled 2×4 MZ-DI device is shown in Fig. 5.10 (a). The measurements were carried out by use of input 1 (i1). The devices were realized in 4 µm SOI substrates using the Omega design.

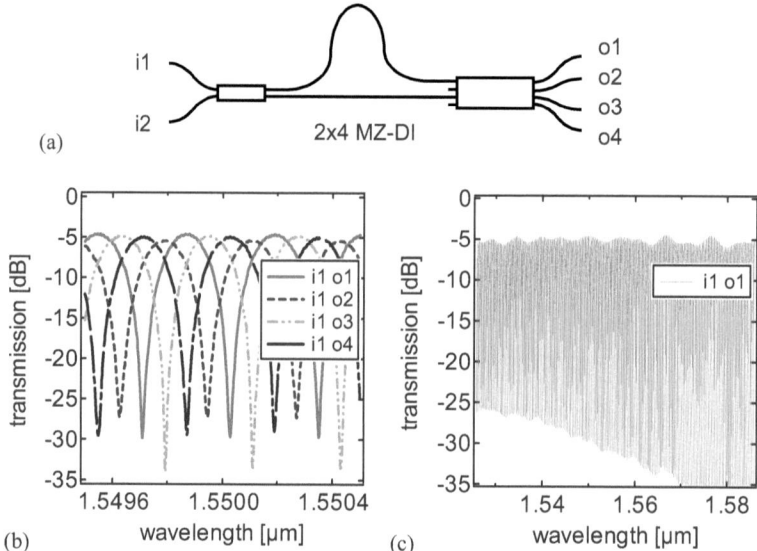

Fig. 5.10 Labeling of the 2×4 MZ-DI. The transmissions of the 2×4 MZ-DI around 1550 nm (all outputs) and other C-band (output 1) for TE polarized light are shown in (b) and (c), respectively.

Measurements for the TE mode around 1550 nm are shown in Fig. 5.10 (b). The transmissions are normalised to straight waveguides from the same chip.

The transmissions in Fig. 5.10 (a) include an intrinsic loss of 3 dB caused by the use of only two input ports (1 and 3) of the 4×4 MMI coupler. A slightly higher imbalance (~ 0.8 dB) compared to 2×2 MZ-DI can be observed. For instance, Fig. 5.10 (b) shows the transmission of output 1 (o1) over C-band. The corresponding extinction ratios of all the outputs are depicted in Fig. 5.11 at temperatures of 35°C and 70°C. Input 1 of the 2×4 MZ-DI was used for the measurements.

The MZ-DI characteristics are very similar at 35°C and 70°C. The extinction ratios remain above 20 dB at all output ports. For balanced detection of the 2×4 MZ-DI output signals, the π-shifted output ports o1 & o4 and o2 & o3 has to feed into the pair of balanced photo diodes.

The imbalances of these output combinations are therefore an important criterion. The results for PDL and IMB over the C-band are shown in Fig. 5.12.

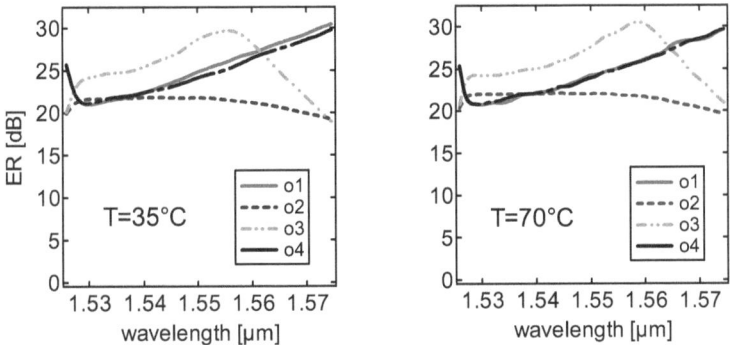

Fig. 5.11 Extinction ratios at output ports of 2×4 MZ-DI at 35°C (left) and 70°C (right) for TE-mode.

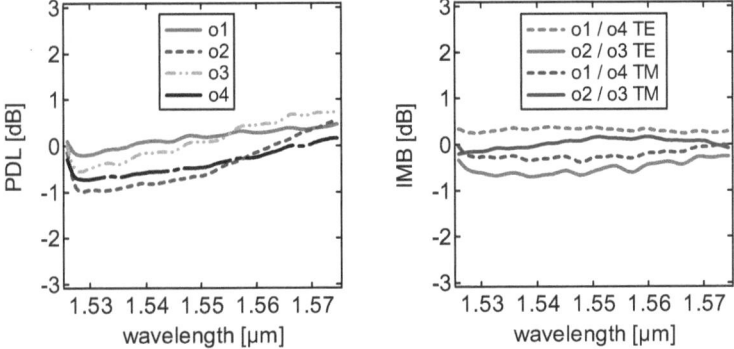

Fig. 5.12 Polarization dependent loss (a) and imbalance (b) of a 2×4 MZ-DI.

Around 1550 nm PDL and IMB do not exceed beyond 0.7 dB. The same result was achieved by the C-band operation, except the slightly higher PDL for the output 3. Finally, the MZ-DI results are similar to the measurements on single 4×4 MMI couplers.

Relative Phases

The relative phases of a 2×4 MZ-DI are shown in Fig. 5.13. The difference from the ideal phase is in the range of ± 5° at each output port. At higher temperature (70°C) a slightly higher phase variation was observed.

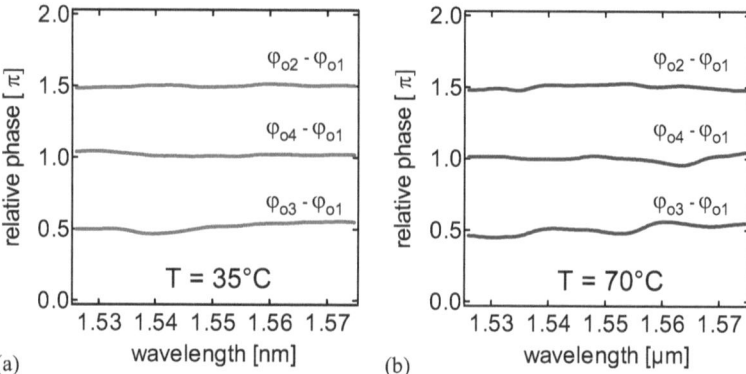

Fig. 5.13 Relative phases at the output ports of a 2×4 MZ-DI at 35°C (a) and 70°C (b). Phase reference for the calculation is the phase at port 1.

The relative phase is calculated with reference to the wavelength at minimum transmission at output 1 by: $\varphi_2 - \varphi_1 = 2\pi \cdot (\lambda_{min.Po2} - \lambda_{min.Po1})/\Delta\lambda_{FSR}$, $\varphi_3 - \varphi_1 = 2\pi \cdot (\lambda_{min.Po3} - \lambda_{min.Po1})/\Delta\lambda_{FSR}$, $\varphi_4 - \varphi_1 = 2\pi \cdot (\lambda_{min.Po4} - \lambda_{min.Po1})/\Delta\lambda_{FSR}$. Here, $\lambda_{min.Po1}$, $\lambda_{min.Po2}$, $\lambda_{min.Po3}$ and $\lambda_{min.Po4}$ denote the wavelength of the transmission minima at output port 1, 2, 3 and 4, respectively. The expression $\Delta\lambda_{FSR}$ corresponds to the wavelength difference of a free-spectral range.

PDFS measurement

The PDF shifts of a 2×4 MZ-DI were determined across the C-band. As shown in Fig. 5.14 (a), the PDF shifts stay below 1 GHz. These shift values exceed the PDFS specification for DQPSK demodulation (~ 0.2 GHz) [14]. It might be an option to use the 2×4 MZ-DI in regions with low PDF shifts. However, this leads to significant bandwidth limitation.

As presented for the 2×2 MZ-DI, polarization dependent measurements were also carried out with linear polarized light in steps of 10° between TE and TM polarized light. A PDFS of up to 3 GHz (± 1.5 GHz) was determined (see Fig. 5.14 (b)).

The PDFS calculation based on the consideration of the maximum shift at the transmission minima.

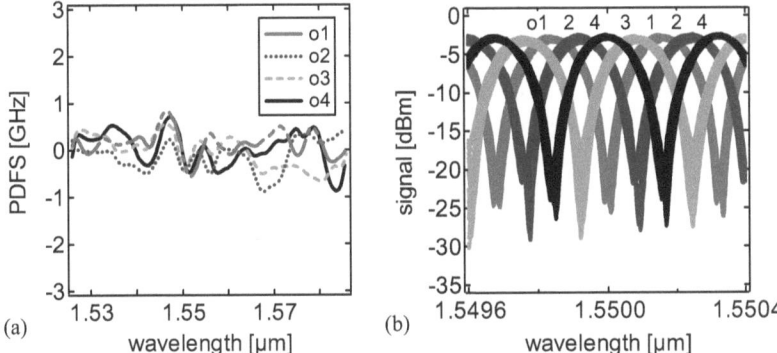

Fig. 5.14 PDFS measurements on 2×4 MZI with TE- and TM polarized light (a). Transmission measurements with linear polarized light in steps of 10° between TE and TM light (b).

5.4 Thermal tuning of MZ-DI

The control of the phase in a MZ-DI can be achieved by means of the thermo-optic effect (TOE), i.e. by heating of one interferometer arm. The heating modifies the phase by variation of the propagation constant. The phase shift can be calculated as follows:

$$\Delta\varphi = \frac{2\pi}{\lambda}\left(\frac{dn}{dT}\right) L_H \Delta T \quad (5.1)$$

Here ΔT denotes temperature difference by heating one interferometer arm with the heater length L_H. The thermo-optic coefficient (TOC) is given as derivation of dn/dT and amounts to ~$1.8 \cdot 10^{-4}$ K^{-1} for silicon [48]. Besides phase variation by thermo-optic effect, the complete phase variation in a MZ-DI has to consider the MZ delay length (ΔL) and the thermal expansion effect. The latter effect implies a phase shift by a differential heater length given by: $\Delta L_H = \gamma_{th} L_H \Delta T$. Here, the thermal expansion is described by the thermal expansion coefficient (TEC) and labeled by γ_{th}. The TEC for silicon amounts to ~$2.5 \cdot 10^{-6}$ K^{-1} [49]. This leads to the following expression for the complete MZ-DI phase variation given by the MZ delay length, the thermal expansion effect and the thermo-optic effect:

$$\Delta\varphi = \underbrace{\frac{2\pi}{\lambda} n_{\text{eff,arm}} \Delta L}_{\text{MZ-delay}} + \underbrace{\frac{2\pi}{\lambda} n_{\text{eff,arm}} \gamma_{\text{th}} L_H \Delta T}_{\text{thermal expansion}} + \underbrace{\frac{2\pi}{\lambda}\left(\frac{dn}{dT}\right) L_H \Delta T}_{\text{thermo-optic}} \quad (5.2)$$

A tuning of MZ-DIs is also feasible by heating of the whole device. Using eq. (5.1) and $L_H = \Delta L \sim 2\text{mm}$, we obtain a π-phase shift for a change of the chip temperature of 2.1 K. Here, the delay length $\Delta L \sim 2$ mm corresponds to a time delay of 25 ps for a 40 Gb/s DPSK demodulator. However, the heating of a complete SOI chips is only adequate at the stage of tuning tests of the MZ-DI devices.

We need to consider the power dissipation leading to a π phase shift by the use of heaters. In case of SOI, the substrate acts as a perfect heat sink due to the high thermal conductivity of silicon ($\sigma_{\text{th,Si}}=150\text{W/(mK)}$, $\sigma_{\text{th,SiO2}}=1.4\text{W/(mK)}$).

In addition, heat is also effectively spread beyond the interferometer arm. Therefore, a considerable amount of power is needed for MZ-DI tuning despite the high TOC of silicon. A thermal resistance model of Fischer [8; 50] similar to electrical networks can be used to estimate the required power for thermal tuning of SOI waveguides. This resistance model describes the lateral heat propagation by a thermal conductance value G'_{th} (length-dependent conductance to the substrate) and R'_{th} (length-dependent thermal resistance along the lateral rib waveguide region. Fig. 5.15 shows a schema of the resistance model [50]. The slab height of the SOI waveguide is defined by h. The thickness of the silicon oxide is given by H_{SiO2}.

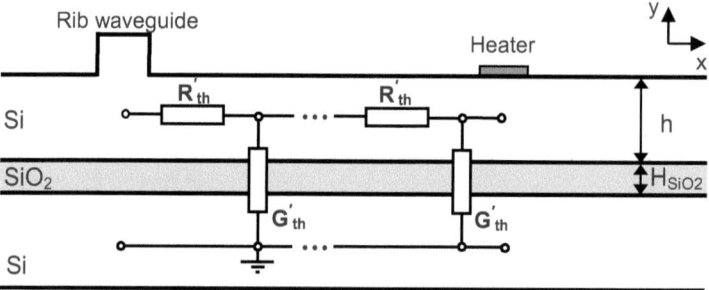

Fig. 5.15 Thermal resistance model of lateral heat propagation in SOI material [50].

The calculation of a thermal diffusion length L_{th} and a thermal decay constant α_{th} can be done as follows [50]:

$$L_{th} \equiv \alpha_{th}^{-1} = \left(\sqrt{G'_{th} R'_{th}}\right)^{-1} \quad (5.3)$$

The length-dependent conductance (G'_{th}) and length-dependent resistance (R'_{th}) can be expressed by:

$$R'_{th} = \frac{\sigma_{th,SiO2}}{h\, L_H} \quad (5.4)$$

$$G'_{th} = \frac{\sigma_{th,Si} L_H}{H_{SiO2}} \quad (5.5)$$

The thermal diffusion can be calculated by the thermal conductivities of silicon $\sigma_{th,Si}$ and silicon oxide $\sigma_{th,SiO2}$ as well as the buried oxide thickness H_{SiO2} and the cladding layer thickness h of the SOI rib waveguide:

$$L_{th} = \sqrt{\frac{\sigma_{th,Si}}{\sigma_{th,SiO2}} H_{SiO2} h} \approx 10 \sqrt{H_{SiO2}\, h} \quad (5.6)$$

The power required to achieve a π phase shift is:

$$P_\pi = \frac{\lambda\, W_{H,eff}}{2\, H_{SiO_2}} \left(\frac{dn}{dT}\right)^{-1} \sigma_{th,SiO_2} \quad (5.7)$$

Here, the effective heater width $W_{eff,heat}$ is given by:

$$W_{H,eff} = 2 L_{th} + W_H \approx 20 \sqrt{H_{SiO_2} h} + W_{heater} \quad (5.8)$$

The effective width $W_{H,eff}$ consists of the thermal diffusion length L_{th} and the chosen width of the thin film heater W_H. The design of the integrated heaters in this PhD thesis was chosen as follows: a heater width $W_H = 10$ μm at h = 2 μm and $L_H =$ 3 mm. With H_{SiO2}=1 μm and following the model of Fischer, the required power P_π with eq. (5.7) amounts to 163 mW.

Indeed, we should expect a slightly higher power dissipation because of the additional lateral displacement of the heater with respect to the waveguide. Direct placement of the heater metal on top of the waveguide is critical due additional losses induced by the metal.

Experimental results

Thermal MZ-DI tuning experiments at TU Berlin followed two approaches:

1) Heating of the complete SOI-chip, and

2) Integrated heaters in cooperation with Fraunhofer Institut Berlin (HHI).

The heat can be generated by a Peltier element, which is inserted in the sample stage and monitored via a resistance temperature detector (PT100). This approach is very useful because no additional fabrication step is required. Fig. 5.16 shows the sample stage at the TU-Berlin with a mounted Peltier element.

Fig. 5.16 Sample stage at the TU-Berlin with inserted Peltier element. The SOI chip is in- and out coupled with lensed SM fibers.

Fig. 5.17 (a) shows the minimum of a MZ-DI transmission as the chip temperature goes from 35°C to 38°C. The MZ-DI transmission minimum moves during the heating to higher wavelengths. Fig. 5.17 (b) shows the experimental wavelength shift at selected temperatures. A temperature change of ~ 2.1 K is required for a π-shift of the 2×2 MZ-DI.

The realization of integrated heaters requires two additional masks. To avoid optical loss, caused by interaction between guided light and heater metal, the heater stripe is not directly placed on top of the rib waveguide. A distance of 15 µm was chosen between rib waveguide and heater stripe.

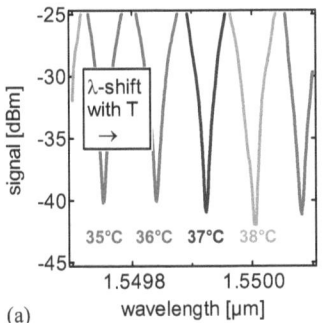 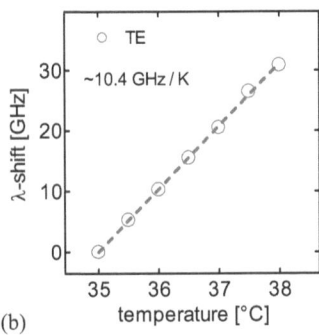

Fig. 5.17 λ-shift of a MZI transmission minimum with increased chip temperature (a). The wavelength shift for TE polarized light is plotted over temperature in (b). The results for TM polarized light are very similar.

No additional optical loss could be observed at this distance. In cooperation with Fraunhofer Institut Berlin (HHI) the following heater dimensions were used: W_H = 10 μm, L_H = 3 mm, material = platinum. The resistance of realized heaters is about 600 Ω. Fig. 5.18 (a, b) show the realized integrated heaters at one interferometer arm of a 50° MZ-DI.

Fig. 5.19 shows the relation between applied power on the integrated heaters and the phase shift in a transmission spectrum. To obtain a π-shift at a 50° MZ-DI and at a Ufo MZ-DI a heater power of ~170 mW and 300 mW is required, respectively. Therefore, the predicted power requirement of Fischer (see above) is verified in the case of the 50° MZ-DI.

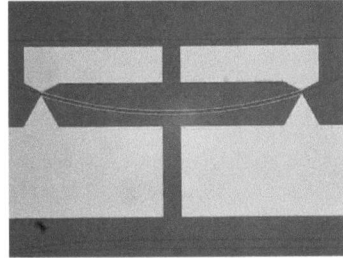 

Fig. 5.18 View of fabricated platinum heaters large-scaled (left) and zoomed (right). The contact areas are deposited with gold layer.

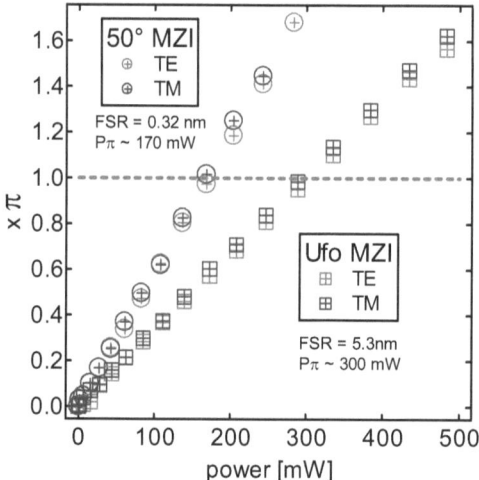

Fig. 5.19 Characteristics of integrated heaters used in 50°MZ-DI and Ufo MZ-DI. The required power for π-shift is increased for Ufo MZ-DI due to higher thermal crosstalk.

The difference in the power consumption results from a larger FSR (5.3 nm) in case of the Ufo MZ-DI. The corresponding delay length is only ~ 100 μm (compared to ~ 2 mm by 50° MZ-DI) leading to a more compact MZ-DI device but also to higher thermal crosstalk between the interferometer arms.

A power reduction could be achieved by the use of grooves around the heaters for thermal isolation. Grooves will normally be fabricated by use of etching techniques (wet or dry). This topic is a current research activity at TU-Berlin.

Another method consists in the laser beam cutting method. A laser beam cuts deep grooves into a material or intersects it completely. First tests on 4μm SOI material were carried out at the Fraunhofer Institut Berlin (HHI).

However, after cutting optical measurements on the MZ-DI devices showed performance degradation (higher insertion loss, low ERs). The reason may be related to the apparent depositions on the SOI chip surface or damages of the rib waveguides. Therefore, further investigation will be required.

5.5 Impacts on MZ-DI birefringence

DPSK demodulators are devices with extremely high sensitivity to birefringence. Therefore, we should also study the effect of process variations on geometrical birefringence. The change of the geometrical birefringence can be caused by a width variation ΔW, an etch depth variation Δ(H-h) and a substrate variation ΔH. Fig. 3.14 (a, b) is already suitable to estimate the related birefringence change. However, the large scaling makes it difficult to analyze small width or etch depth variation. Fig. 5.20 (a, b) shows the same results in a smaller range of the rib waveguide width W and the etch depth (H-h).

The change in birefringence by variation of nominal rib waveguide width W = 3.5 μm is relatively small. Furthermore, the change in birefringence given by a variation in rib height H is moderate due to the insensitivity around W = 3.5 μm. The largest change in birefringence is given by etch depth variation. The results are collected in Tab. 5.1.

Here, the change in birefringence is expressed as ΔPDFS by: $f \cdot \Delta(\Delta n_{geo})/n_g$. Here, f denotes the frequency of light, $\Delta(\Delta n_{geo})$ the change of geometrical birefringence and n_g the group index of a standard SOI rib waveguide (~3.61). In real MZI devices made from the same 4μm SOI wafer were found a PDFS variation of about 3 GHz. The realization of the devices with uniformly low PDFS may therefore require an extended birefringence tuning on single devices.

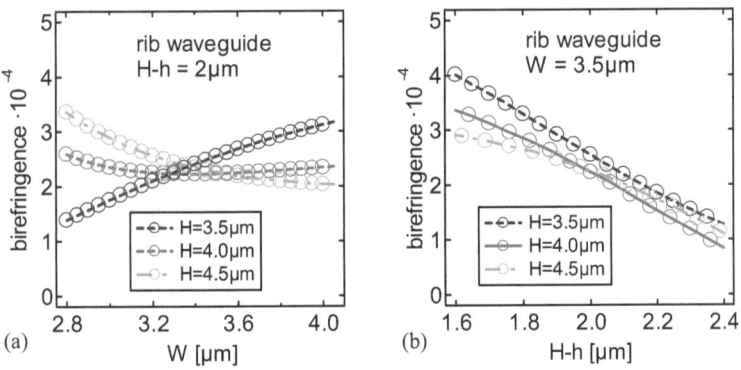

Fig. 5.20 Variation of geometrical birefringence Δn_{geo} as function of rib waveguide width (a) and etch depth (b). Additionally, the influence of substrate thickness variation is included.

Tab. 5.1 Change in PDFS resulting from process variations and substrate non-uniformity.

Impact	Variation [µm]	Δ PDFS [GHz]
ΔW	±0.2	±0.3
ΔED	+0.3	-5.3
ΔH	±0.5	±1.6

Applications on a 4μm SOI material platform

6 Hybrid integrated DPSK receiver

6.1 Background

Over many years optical communication systems were dominated by use of conventional on-off keyed (OOK) signals, i.e. intensity modulated signals. During the last decade, a number of advanced formats (beyond OOK [9]) for high speed transmission applications have been studied. Beside amplitude, information is carried by the optical phase (phase shift keying, PSK). More specifically, to avoid the need for an absolute phase reference, differential-PSK (DPSK) deploys the optical phase change between preceding bits. Compared to OOK, the balanced detection of DPSK signals provides a ~3 dB benefit in terms of average optical power leading to higher receiver sensitivity or lower OSNR requirements, respectively. That can be exploited to extend transmission distances, reduced optical power requirements or relaxed component specifications [10].

The work on a hybrid integrated DPSK receiver was performed in a cooperation of TU-Berlin with the Fraunhofer Institut Berlin (HHI) and u2t-photonics AG in the project EIBONE (Efficient Integrated Backbone), which was funded by the German BMBF. Fig. 6.1 shows the composition of the hybrid integrated DPSK receiver (a) as well as a photograph of a realized receiver (b).

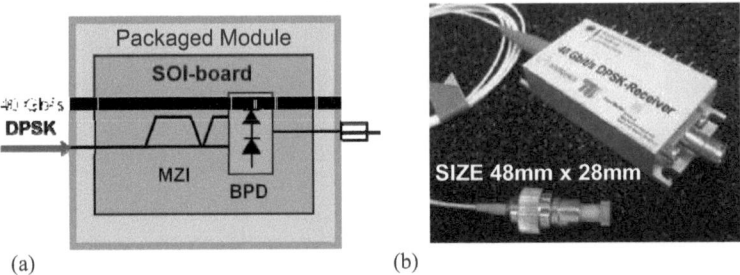

Fig. 6.1 Composition of a hybrid integrated DPSK-receiver (a) and the photograph of a realized integrated receiver (b).

In that configuration, an optical 2×2 MZI for demodulation of 40 Gb/s signals is followed by a balanced photo detector (BPD). The BPD is mounted on a SOI-board via gold-tin flip-chip soldering. Finally, the SOI-board is inserted into a package with ports for external connection (DC, RF, optical fiber).

6.2 Layout and Flip-chip process

Fig. 6.2 shows the 4µm SOI board layout for DPSK signal demodulator with chip size of 26 mm × 13 mm. The board layout contains three equal 50° MZ-DIs. The 50° MZ-DI layout was used because of the space-saving arrangement of MMI couplers and interferometer arms compared to the Omega or Pi MZ-DI layout. Two of the MZ-DIs are connected with an integration zone for a BPD integration at the output waveguides. In the final package only one MZ-DI is supported, therefore the second 50° MZ-DI provides device redundancy. A third MZ-DI serves as test MZ-DI for birefringence control on chip and can be coupled to lensed SM fiber (after polishing).

Various waveguide test structures are arranged on top and at the bottom of the layout to enable MZ-DI independent waveguide characterization. The SOI board waveguides are tapered in front of the BPD to achieve low optical coupling losses to the BPD. Fig. 6.3 shows the complete wafer layout (4 inch) with 6 single MZ-DI chips. On top and bottom are chips with test structures to optimize the hybrid integration of balanced photo detectors. The wafer layout was carried out by T. Mitze. The fabrication of SOI boards for flip-chip integration of photo detectors requires a substantial number of additional process steps. The board fabrication process is based on the work of T. Mitze (TU-Berlin) [2] and J. Kreissl (Fraunhofer Institut Berlin (HHI). A scheme of the SOI board for flip-chip mounting of a photo detector is shown in Fig. 6.4.

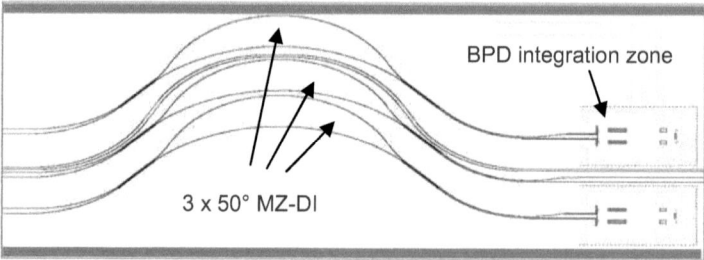

Fig. 6.2 Designed MZ-DIs for DPSK-demodulation. Adjacent to the 50° MZ-DI is the integration zone for a balanced photo detector (BPD) and corresponding contact pads.

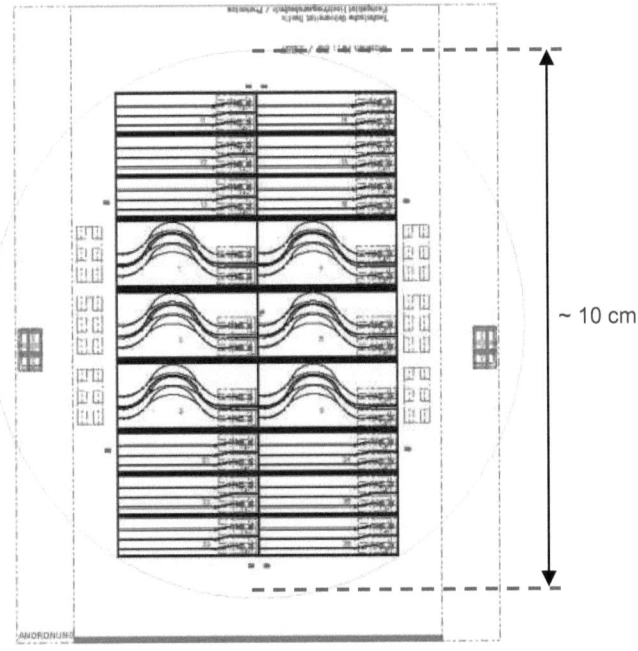

Fig. 6.3 Complete DPSK demodulator wafer layout. Test structures for hybrid integration are arranged on top and bottom.

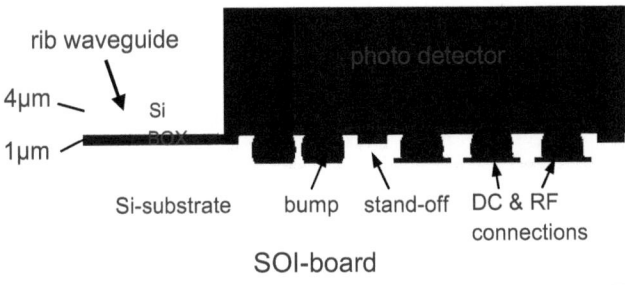

Fig. 6.4 Scheme of the integration zone on a SOI board for flip-chip mounting of a photo detector.

The motherboard fabrication starts with the realization of rib waveguides as described before (4μm SOI rib waveguide technology). A second etch step defines the buried oxide (BOX) layer as vertical adjustment plane for the photo detector (stand-offs) and defines the waveguide facets. A third etch (~ 15μm) defines the integration zone, i.e. the plane for gold-tin solder bumps and electrical connections. The following lithographic steps define the required metal patterns on the chip (fanout, heater, under-bump metallization). The final level realizes AuSn solder studs by sputter deposition. During a reflow process the AuSn retracts from the non-wetting metal and the desired sphere-shaped bump is formed. That is important to ensure a good contact between bumps and contact pads of the active device. The RF connection is implemented by coplanar transmission (CPW) lines. After device separation by dicing and facet polishing, the rib waveguide birefringence can be fine-tuned by cladding induced stress. An anti-reflection coating is applied to reduce coupling loss.

High-speed photo detectors for DPSK receivers can be realized in InP-technology. At Fraunhofer Institut Berlin (HHI) stand-alone BPDs achieve bandwidths of 80 GHz and beyond [51]. The chip surface contains mechanical structures (e.g. standoffs) for the adjustment on a SOI board. The coplanar RF interconnection on-chip is based on 50 Ohm design. An anti-reflection coating is applied on BPD chips with central wavelength at 1550 nm. Fig. 6.5 (b) shows a microscope picture of a BPD chip (bottom side).

The optical coupling from the SOI board to the photo detector chip is supported by a monolithic integrated taper, which is optimized to match spot-size of an optical fiber (SMF 28). Hence, a rib waveguide geometry has to be found with minimum coupling loss to a standard optical fiber.

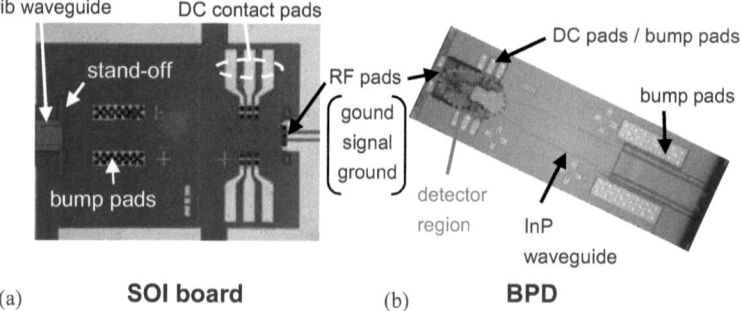

Fig. 6.5 Top view on the integration zone of 4μm SOI-board (a). Microscope picture of a BPD from bottom side (b).

At fixed substrate height and etch depth (H = 4 µm, H-h = 2 µm), the remaining free parameter is the rib waveguide width. Calculations show minimum coupling loss at a rib waveguide width W of ~ 13.5 µm. Fig. 6.6 show the result of the calculations for TE and TM mode. The minimum coupling loss (no misalignment in x, y, z direction) at a rib waveguide width of 13.5 µm is ~ 3 dB for TE mode and ~ 3.2 dB for TM mode. An anti-reflection layer was used during the calculations.

In the final design a rib waveguide width of 14 µm was used. Fig. 6.7 (a) shows a SEM picture of a realized rib waveguide for coupling to the BPD.

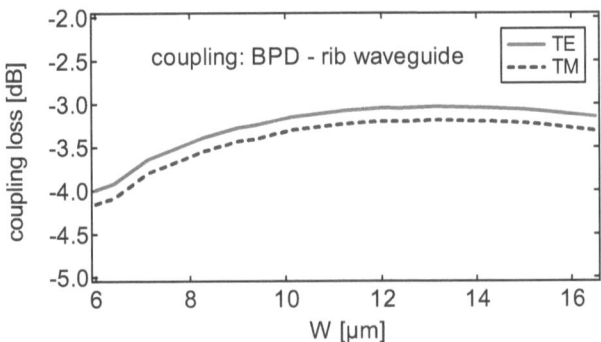

Fig. 6.6 Calculated coupling loss between balanced photo detector (BPD) and a 4µm SOI rib waveguide as function of rib waveguide width W.

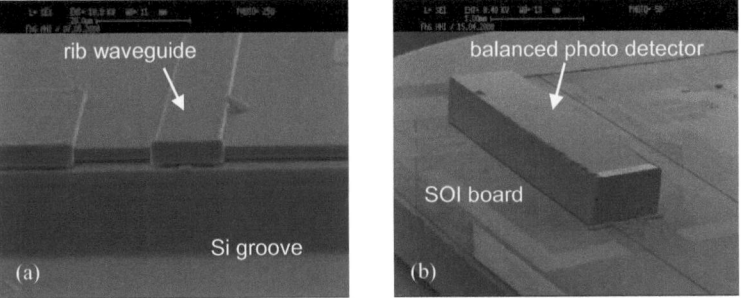

Fig. 6.7 SOI rib waveguide in front of the integration zone for coupling to balanced photo detector (a). Figure (b) shows an integrated balanced photo detector on SOI board.

Placements with a precision of better than 1 μm are nominally provided by the flip-chip tool. Suitable temperature ramps are also required for high-quality soldering. Fig. 6.7 (b) shows a SEM picture of an integrated BPD on SOI board.

Furthermore, the coupling loss is very sensitive to misalignment in x, y and z-direction. Misalignments lead to additional coupling loss as shown in Fig. 6.8 (a, b). Here, the coupling loss for TE mode is calculated for misalignment in x (horizontal), y (vertical) and z (longitudinal) direction. The results are similar for TM polarized light. However, considering the small tolerances of the etch depth in the board fabrication, we may expect a misalignment in horizontal direction rather than in vertical direction. As an example, a misalignment of $x = 2$ μm and $z = 2$μm (at the same time) results in an additional loss of 2.64 dB. For calculation of the total coupling loss, a fundamental loss of 3.01 dB at optimum alignment (no misalignment in x,y,z direction) has to be added. The coupling loss sensitivity as function of x,y-misalignment decreases with higher distances z.

The package of the integrated receiver has a footprint of 48 mm × 28 mm. Finally, the following components need to be accommodated: the SOI board on a submount and a Peltier element for thermal adjustment of the SOI board. Furthermore, the package contains an optical PC (physical contact) and a RF connector as well as pins for power supply of the integrated component and control of the Peltier element.

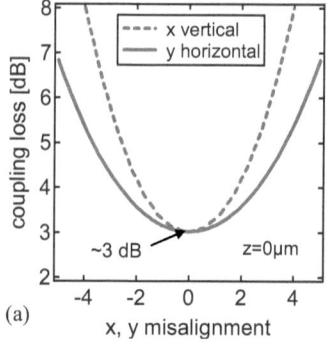
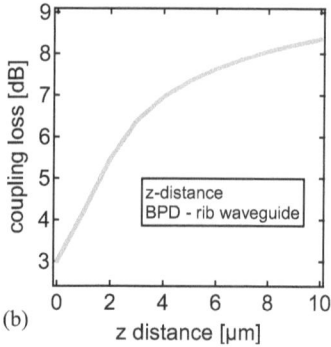

Fig. 6.8 (a) shows calculated coupling loss between BPD and SOI rib waveguide (W = 14 μm, H = 4.0 μm, H-h = 2.0 μm) as function of horizontal (x) and vertical (y) misalignment by use of TE polarized light (z = 0μm). The results for TM polarized light are similar. The z-dependency on coupling loss at optimum x- and y-alignment (x, y = 0μm) is shown in (b). An anti-reflection layer is used during the calculations.

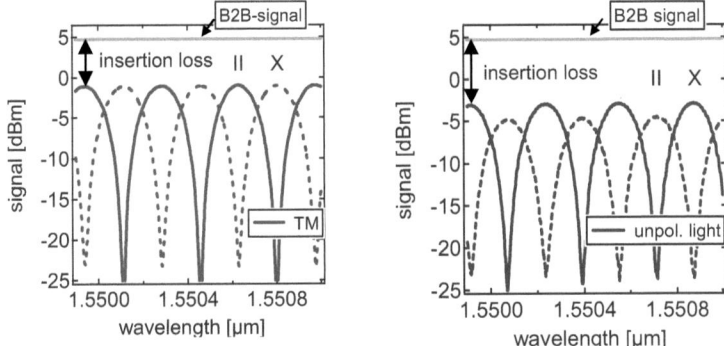

Fig. 6.9 Transmission signals of the same 50° MZ-DI non-packaged (lab stage) and packaged around 1550 nm (a, b). Corning fibers were used in both constellations for coupling of the MZ-DI in- and outputs.

The performance of a packaged MZ-DI device can be degraded due to stress (induced by fixation of the SOI chip) or higher coupling loss between lensed fibers and SOI chip. Therefore, first tests investigated the MZ-DI performance of SOI boards after packaging. Here, the coupling on MZ-DI in- and output waveguides based on lensed fibers, i.e. at this stage no integrated BPD was used. For comparison, Fig. 6.9 shows the transmissions of the same MZ-DI before and after the packaging. The chip packaging increases the MZ-DI imbalance and insertion loss. The imbalance rises from < 0.3 dB up to 1.6 dB. The insertion loss is increased by 2.2 dB (5.8 dB to 8 dB). Both effects can be attributed to various coupling between lensed fibers and SOI chip, which is highly sensitive to process-related misalignment (fiber positioning, glue shrinkage). As shown in simulation, the misalignment of ~ 1 µm leads to an insertion loss > 1dB. The extinction ratios are slightly decreased.

The PDFS of the packaged MZ-DI was about 2 GHz (before packaging 0.5 GHz). This increase is probably a consequence by the fixation of the SOI chip in the package influencing the beforehand adjusted birefringence. However, the package reasonably retains the basic MZ-DI characteristics.

6.3 System Performance of a 40 Gb/s DPSK demodulator

The system performance of the 40 Gb/s DPSK receiver was initially tested without integrated photo detector, i.e. the bare SOI chip (non-packaged) with MZ-DI was investigated.

Here, the balanced photo detector was connected via lensed SM fibers with the MZ-DI outputs. The measurements were carried out at Fraunhofer Institut Berlin (HHI). The setup is schematically shown in Fig. 6.10.

The CW-signal from an external cavity laser was DPSK-modulated using a dual-drive Mach-Zehnder modulator, operating in push-pull mode. The electrical signals to drive the modulator were pseudo random bit sequences with a word length of 2^7-1, amplified in high-bandwidth broadband driver amplifiers. The RZ modulation format was generated with an EAM-based pulse carver (duty cycle 40-50 %). The modulated 40 Gb/s signal passed an attenuator, was amplified, fiber coupled in and out of the 2×2 MZ-DI, and detected by an integrated balanced photo detector (BPD).

Fig. 6.10 shows additionally the point of measurements for received power and optical signal-to-noise ratio (OSNR). The SOI MZ-DI was temperature-controlled during the measurements.

Fig. 6.11 shows the eye diagrams after demodulation and balanced detection, measured with a digital scope including a 70 GHz sampling head (NRZ measurement was done without EAM). For RZ-DPSK, the bit-error rate (BER) was measured as function of received power.

The resulting characteristics are plotted in Fig. 6.12 (a, b). To demonstrate the dependence of BER performance on PDFS, we compare two 2×2 MZ-DI devices: MZ-DI-1 (PDFS: 0.4 GHz), and a device slightly off the optimum birefringence, MZ-DI-2 (PDFS: 2.5 GHz). The input polarization was varied by means of a looped-fiber polarization controller. The state of polarization (SOP) in Fig. 6.12 (a) was labelled *best case* for optimum BER performance, and *worst case* for lowest BER performance.

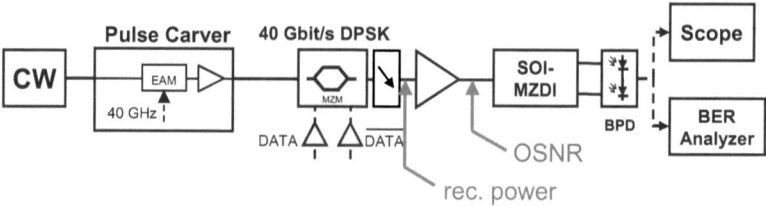

Fig. 6.10 40 Gb/s DPSK setup to test the demodulation performance of the 2×2 MZ-DI. The received power and the OSNR were determined before and after the amplification of the DPSK signal, respectively.

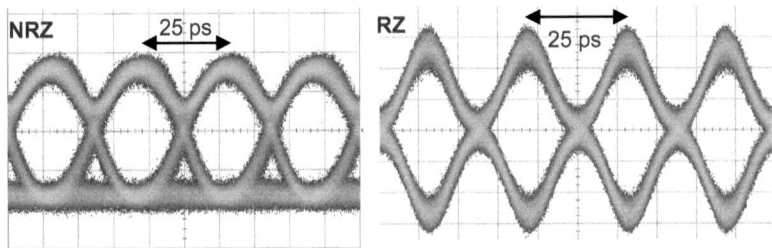

Fig. 6.11 Eye diagrams after demodulation of 40 Gbit/s NRZ/RZ and balanced detection measurements (a, b).

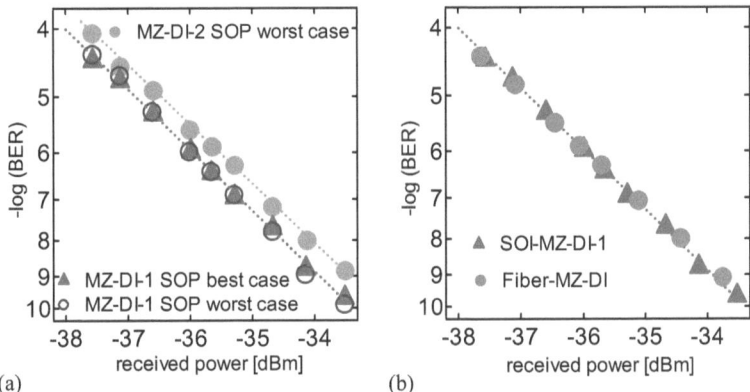

Fig. 6.12 (a) BER performance for best & worst case input polarization for MZ-DI-1 (PDFS: 0.4 GHz), and for MZ-DI-2 (PDFS: 2.5 GHz). The polarization dependent penalty is only significant in case of MZ-DI-2 (~ 1 dB). (b) BER performance of MZ-DI-1 in polarization optimum and a fiber delay interferometer at a bit rate of 40 Gb/s. The performance of the two technologies is comparable.

The measurement for BER performance in Fig. 6.12 (a) shows a polarization dependent penalty of 0.1 dB in the case of MZ-DI-1 (small PDFS). For MZ-DI-2 the BER performance is slightly worse due to the larger PDFS of the device. The penalty in BER performance is ~ 1 dB. Finally, a BER of 10^{-9} corresponds to a received power of ~ 34.1 dBm. Fig. 6.12 (b) compares the performance of a 40 Gb/s MZ-DI in 4μm SOI technology with a device in fiber technology for an arbitrary polarized input light. The two BER curves are virtually indistinguishable, demonstrating equivalent performance of the two technologies.

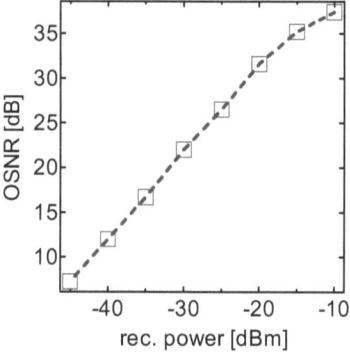

Fig. 6.13 Relation between OSNR and received power.

The 40 Gbit/s DPSK system measurements based on SOI 2×2 MZ-DI show therefore state-of-the-art performance and demonstrates the high potential of SOI rib waveguide technology for low-PDFS applications. The relation between BER performance and OSNR can be determined by a calibration of OSNR versus received power as presented in Fig. 6.13. Thus, the received power of -34.1 dBm corresponds to an OSNR of 18 dB.

6.4 System performance of a 40 Gb/s DPSK receiver module

Following successful MZ-DI performance tests for phase demodulation on chip-level, also system measurements of 40 Gb/s DPSK receiver module with integrated BPD were pursued [52]. The measurements were carried out by u2t photonics with 40 Gb/s non-return-to-zero (NRZ) DPSK signals. Fig. 6.14 shows the test bed for measurements, which is similar to the system test bed for the MZ-DI without integrated BPD.

The measured diagrams of the sampling oscilloscope show a clear eye opening and a bit pattern without degradation due to pattern effects (see Fig. 6.15). The non-amplified signal has an amplitude of 60 mV.

The performance tests resulted in a bit-error rate (BER) performance vs. received power and OSNR as shown in Fig. 6.16 (a, b) for pseudo random bit sequences (PRBS) with a word length of 2^{31-1}. A bit-error rate of $< 10^{-9}$ is achieved for an optical power (P_{rec}) of > -33 dBm. The corresponding OSNR for error-free detection is 20.8 dB. For a PRBS with word length of 2^7-1 error free operation was achieved for an OSNR of 20.4 dB [52].

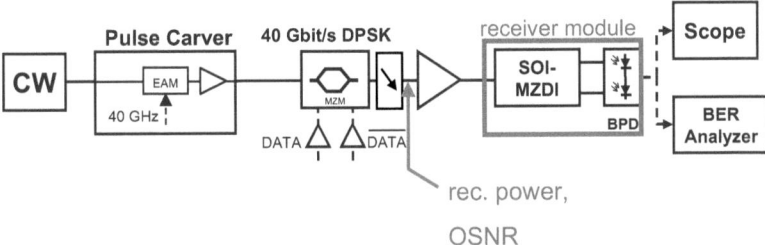

Fig. 6.14 Test-bed for system measurements of 40 Gb/s DPSK receiver. The received power and the OSNR were determined before amplification of the DPSK signal.

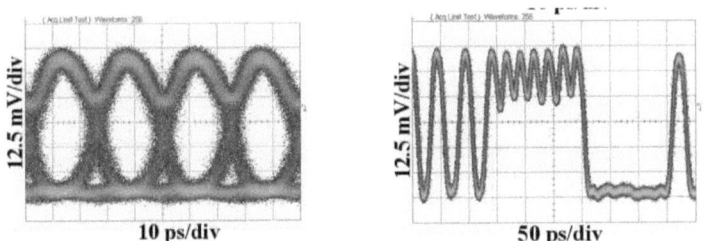

Fig. 6.15 The received eye diagram (left) and the bit pattern diagram (right) of the DPSK receiver in test-bed for 40 Gb/s NRZ signals.

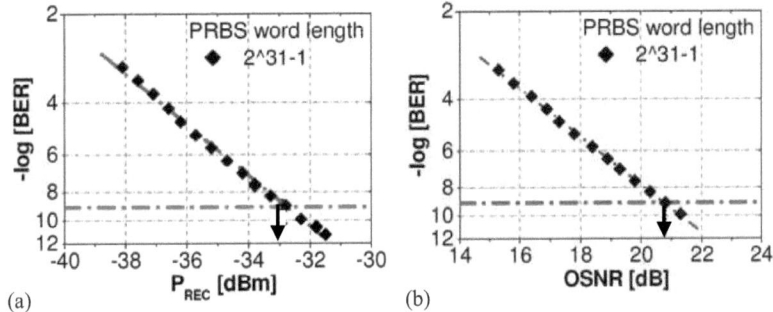

Fig. 6.16 BER performance versus received power (a) and OSNR (b) for pseudo random bit sequences (PRBS) with word length of 2^{31-1}.

7 SOI material as platform for all-optical wavelength conversion

7.1 Introduction

All-optical wavelength conversion is considered as a building block in future high-capacity wavelength-division-multiplexed networks [53]. Recent research activities consider semiconductor-optical amplifiers (SOA) for AOWC by taking advantage of their nonlinearity. In this regard, SOAs provide the potential for ultrafast wavelength conversion with a speed > 100Gb/s. Furthermore, SOAs have a high integration potential.

A number of different SOA-based AOWCs have been demonstrated in the past [54; 55; 56]. Liu showed error-free AOWC operation at a record bit rate of 320 Gb/s [53]. Here, the AOWC is made by the following components: SOA, a fiber Bragg grating, an optical bandpass filter and a fiber-based delayed interferometer. A general description of the AOWC operation principle can be found in [57].

An AOWC as integrated component was so far not presented. The target of the European project "BOOM" is the development of an integrated AOWC based on SOI board technology. The TU-Berlin cooperates in BOOM with the Fraunhofer Institut Berlin (HHI), the National Technical University of Athens (NTUA), the University of Valencia and the Technical University Eindhoven.

This thesis focuses on the work mainly done at TU-Berlin, i.e. SOI board layout, board fabrication (together with the Fraunhofer Institut Berlin (HHI)) and basic optical SOI board characterization.

7.2 SOI-board for wavelength conversion

In the BOOM project, 4µm SOI material serves as a motherboard for SOA hybrid integration and provides two monolithically integrated periodic filters. The passive filter following the SOA consists of two cascaded Mach-Zehnder delay interferometers. Fig. 7.1 shows a schematic of the integrated wavelength converter. The optimum MZ-DI delays for wavelength conversion up to 160 Gb/s were determined at National Technical University of Athens (NTUA). Delays of 1.25 ps (MZ-DI-1) and 2.5 ps (MZ-DI-2) were being selected. The FSRs are thus in the nm-range: ~ 4 nm and ~ 8 nm, respectively.

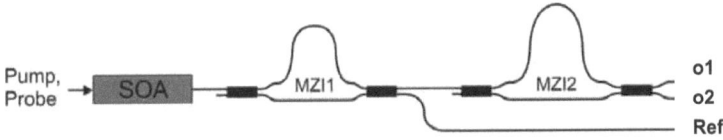

Fig. 7.1 Scheme of the proposed AOWC in the European project BOOM. Beside the outputs of the second MZ-DI, the converter provides one output of the first MZ-DI as reference.

Fig. 7.2 shows the layout of the SOI board (chip-size: 32 mm × 12 mm). Yellow highlighted regions denote metallization areas, i.e. bumps and contact pads for soldering and electrical control of the SOA, respectively. The layout includes also integrated heaters for control of the delay interferometers as well as different test structures. Rib waveguide and heater dimensions correspond to standard values as presented in this thesis. According to package requirements, the distance of output waveguides is 4 mm. The layout provides two SOA integration zones. Therefore, the flip-chip procedure can be carried out twice in case of a failed flip-chip run.

The complete fabrication of the 4µm SOI board includes a set of 8 masks. 10 SOI boards can be realized per wafer. Fig. 7.3 shows the complete wafer layout. The board fabrication was carried out in cooperation with Fraunhofer Institut Berlin (HHI). In detail, TU-Berlin provided the SOI rib waveguides, the grooves for the SOA integration and supported SU-8 (negative resist) lithography. The Fraunhofer Institut Berlin (HHI) was responsible for the metallizations.

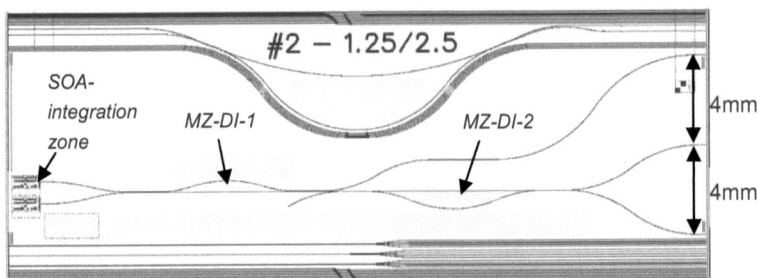

Fig. 7.2 Layout of SOI board for wavelength conversion. The layout starts with an integration zone for a semiconductor optical amplifier (SOA) followed by cascaded Ufo MZ-DIs (MZ-DI-1, MZ-DI-2). Test structures are also included.

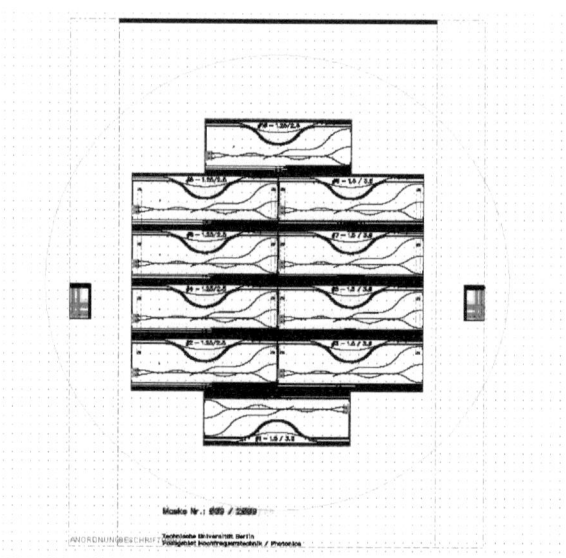

Fig. 7.3 Complete wafer layout of the SOI board. The layout includes 10 single SOI boards.

Fig. 7.4 shows a microscope picture of the integration zone (left side) of a fabricated SOI board for SOA integration. Fabricated soldering bumps after reflow-process are shown in Fig. 7.4 (b).

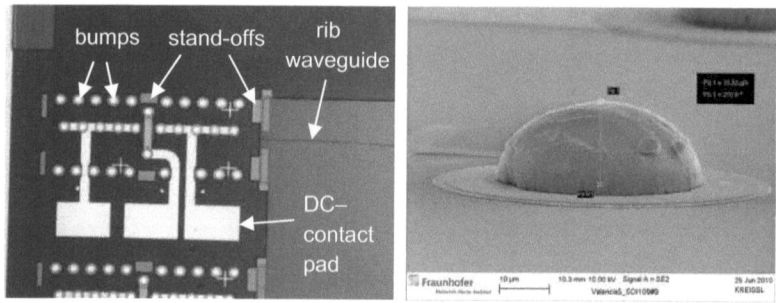

Fig. 7.4 Microscope picture of a fabricated integration zone for a semiconductor optical amplifier (SOA) on 4μm SOI board (left side). SEM picture of sphere-shaped bump after reflow process (right side).

7.3 AOWC with cascaded SOI MZ-DIs

First optical measurements were carried out without an integrated SOA, i.e. only the filter characteristic of the cascaded MZ-DI structure was determined. For that purpose, the SOI chips were polished on both edges to enable low loss fiber-chip coupling. Fig. 7.5 shows the filter characteristics of the cascaded MZ-DI structure with delays of 1.25 ps and 2.5 ps (T = 35°C). Corresponding to an ideal cascaded MZ-DI characteristic, the smaller FSR of the second MZ-DI (delay = 1 ps) splits precisely the FSR of the first MZ-DI (delay = 1 ps). The extinction ratios exceed 25 dB for TE and TM polarization. The signal loss of about 2 dB (reference to S-bend) indicates low additional waveguide loss and 2×2 MMI coupler loss of < 0.5 dB. At higher chip temperatures, no degradation on MZ-DI performance could be observed. Fig. 7.6 (a) shows exemplary the filter characteristics of output 2 at stage temperatures T of 35°C and 85°C. A temperature change of ~ 50°C is necessary to shift the filter curve with one FSR (~ 4nm). The thermal tuning is more specified in Fig. 7.6 (b). As shown for single MZ-DIs (see Fig. 5.17), the wavelength shift is about 10 GHz/K (80 pm/K).

The AOWC functionality was experimentally tested at TU-Eindhoven with the setup shown in Fig. 7.7 [58]. Here, the SOA is a stand-alone device not integrated on the SOI board. Optical pulses are generated by a fiber mode-locked laser (FMLL) with duration of 1.2ps at 1556.3 nm and 40 GHz repetition rate. The pulses are amplitude modulated to form a 2^7-1 return-to-zero on-off keying (RZ-OOK) PRBS signal and time-multiplexed to constitute a 160 Gb/s bitstream.

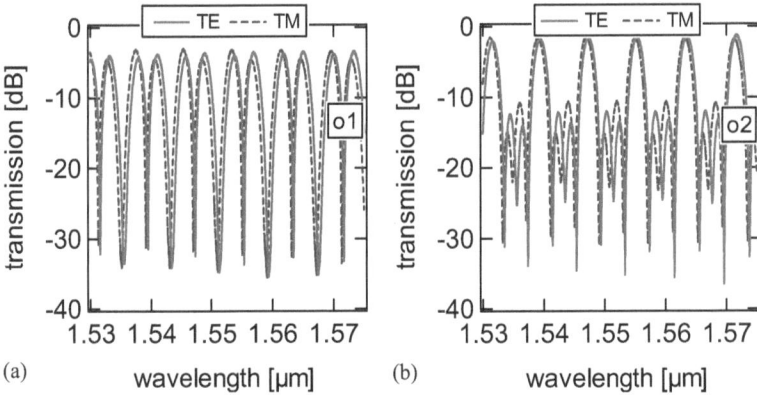

Fig. 7.5 Filter characteristic of cascaded MZ-DI with 1.25 ps and 2.5 ps delay lines for TE (solid line) and TM (dashed line) polarized light. Figure (a) shows the filter characteristic at output 1 (o1) and figure (b) the filter characteristic at output 2 (o2).

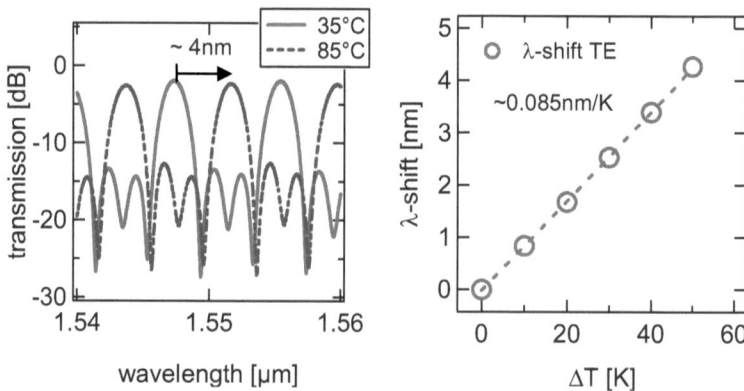

Fig. 7.6 Transmission of cascaded MZ-DI (output 2) at 35°C and 85°C chip temperature for TE polarized light (a). Figure (b) shows the temperature dependent wavelength shift by steps of 10 K.

This signal is coupled into an SOA in combination with CW-light at a wavelength of 1546.6 nm (probe). At the SOA output, an optical band-pass filter (OBF) with a 3 dB bandwidth of 5 nm is used to select the probe and reject the pump signal. Then the signal is coupled into the SOI photonic circuit. The output of the SOI photonic circuit is amplified by a low noise erbium-doped fiber amplifier (EDFA) and filtered for bit-error-rate (BER) performance evaluation.

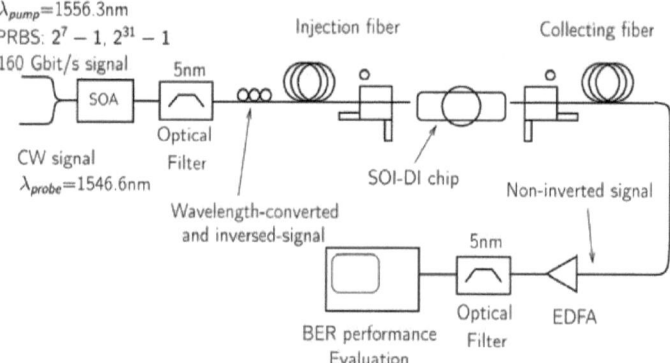

Fig. 7.7 Experimental setup for wavelength conversion at 160 Gb/s [58].

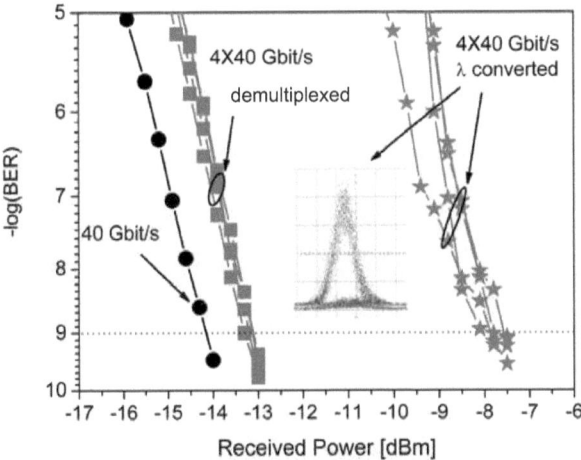

Fig. 7.8 BER curves of the SOA-based AOWC. Circles represent single channel (40 Gb/s) without wavelength conversion; squares represent the degradation of the demultiplexed wavelength converted 4×40 Gb/s channel without wavelength conversion and stars the 4×40 Gb/s channel with wavelength conversion [58].

Fig. 7.8 shows the BER performance of the system. The circles illustrate the error-free operation of a single 40 Gb/s channel that corresponds to the case without wavelength conversion. The squares represent the degradation of the wavelength converted and demultiplexed 4×40 Gb/s channel, whereas the stars represent the wavelength converter channels using the SOA-based AOWC. All the channels achieved error-free operation with an average optical receiver power of -7.5 dBm associated to an average power penalty of 5.5 dB. Finally, the results show successful AOWC operation at 160 Gb/s employing a photonic circuit with cascaded MZ-DIs in 4μm rib waveguide technology.

7.4 AOWC with integrated SOA on SOI board

The hybrid integration of the SOA follows the approach shown for coarse WDM transmitter modules [2]. SOA-devices will be soldered up-side down to enable butt-coupling to the SOI rib waveguide. The vertical SOA alignment based on the use of the oxide plane of the SOI-material (stand-offs). The lateral alignment is carried out by a flip-chip bonder, which provides positioning accuracy of ≤ 1 μm.

The flip-chip soldering was developed in collaboration with the packaging platform ePIXpack (Mr. G. B. Preve, University of Valencia). Fig. 7.9 shows a fabricated SOI board with integration zone (left side) and a flip-chip soldered SOA (right side). Due to optimizations, the quality and accuracy of the flip-chip process could be clearly improved from May to December 2010. One step consisted in the revision of temperature ramps by soldering. Another step was the manufacturing of an adapted chip holder, which respects the SOA positioning near the edge of the SOI board. Fig. 7.10 collects x- and z-misalignment measurements from May to December 2010. Finally, the SOA alignment can now repeatably carried out with tolerances in the x-direction of ± 1 µm.

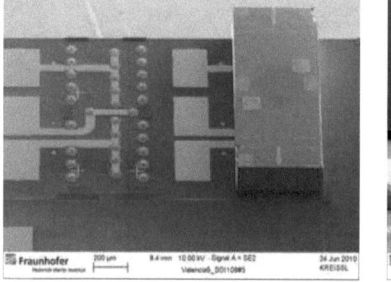 

Fig. 7.9 SEM pictures on integration zone (left side) and integrated SOA (right side) on SOI board.

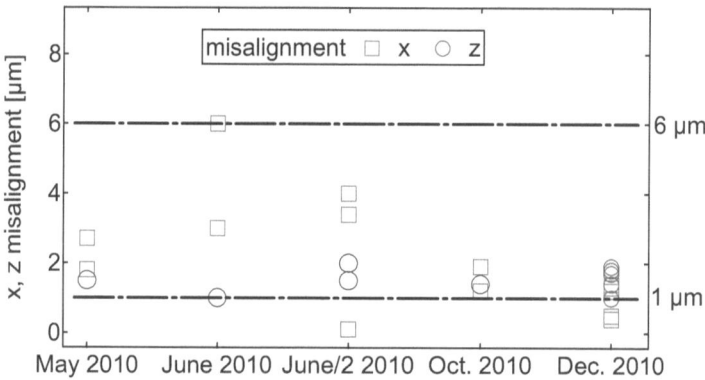

Fig. 7.10 Alignment progress of flip-chip mounted SOAs from May to Dec. 2010.

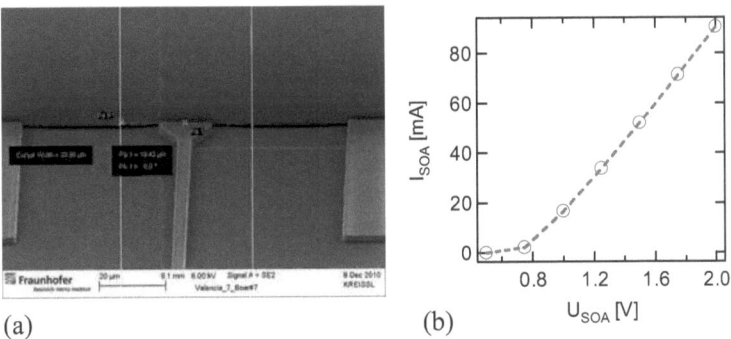

(a) (b)

Fig. 7.11 SEM picture of flip-chipped SOA and rib-waveguide on SOI board (a). The x-misalignment in is about 0.5 µm (Dec. 2010). The U/I characteristic of a SOA is shown in Fig. (b). Applied voltages and currents are denoted by U_{SOA} and I_{SOA}, respectively.

Fig. 7.11 shows exemplary a SEM picture from December 2010. The misalignment in x-direction is 0.5 µm. In the next step, the complete AOWC (integrated SOA plus cascaded MZ-DIs) was optically tested at TU-Berlin. A laser source delivers an input power of P_{laser} = 6 dBm over the C-band. The SOA was driven with current (I_{SOA}) of 18mA and 100 mA, respectively. The related voltage can be determined by the U_{SOA}/I_{SOA} characteristic shown in Fig. 7.11 (b). Fig. 7.12 shows the transmission of the AOWC.

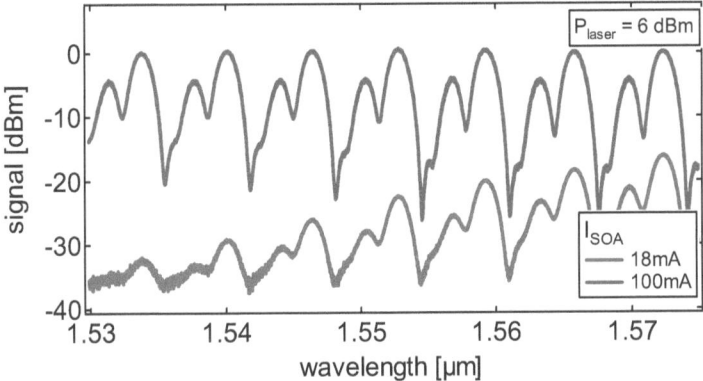

Fig. 7.12 Transmission of the AOWC (integrated SOA + cascaded MZI). The SOA current is 18mA and 100mA.

Corresponding to the diode characteristic of the SOA, the MZ-DI filter curves appears at a SOA driving current of 100 mA. The signal difference between the B2B-signal and the output signal is about 4 dBm. This has to be attributed to fiber coupling loss to the SOA and the rib waveguide facets as well as non-optimum coupling between the SOA and the SOI rib waveguide. Dynamic measurements on devices with an integrated SOA on SOI board were carried out at National Technical University of Athens (NTUA) [59]. The setup is shown in Fig. 7.13.

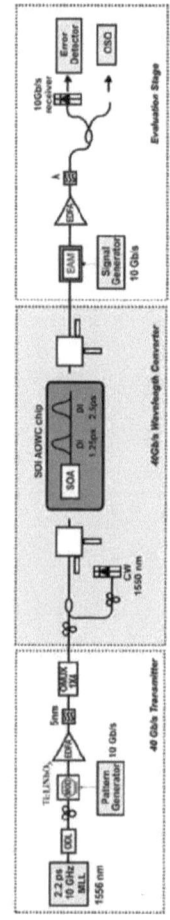

Fig. 7.13 Setup for AOWC measurements at NTU Athens [59].

Fig. 7.14 shows eye-diagrams at a data rate of 40 Gb/s. Fig. 7.14 (a) represents an incoming 40 Gb/s data signal, which was injected into the SOA for wavelength conversion. This signal passes through the SOI chip where chirp filtering and pulse polarity inversion takes place. Fig. 7.14 (b) shows the inverted wavelength converted signal when detuning the MZ-DI filter curves with 0.2 nm off the signal carrier for effective chirp filtering and SOA acceleration. Fig. 7.14 (c) depicts the non-inverted wavelength converted signal. With a total power of 40 mW in the integrated heating elements, a response with 25 dB extinction was obtained for sufficient carrier suppression. Although the signal polarity was effectively restored, very low output power has been measured at the output of the AOWC (-30 dBm) resulting to an amplified waveform with low OSNR (~ 18 dB) and amplitude noise.

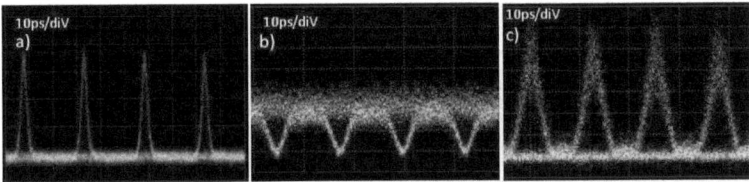

Fig. 7.14 Eye-diagrams of the B2B-signal (a), the inverted signal after MZI1, and the non-inverted signal after (MZI2) [59].

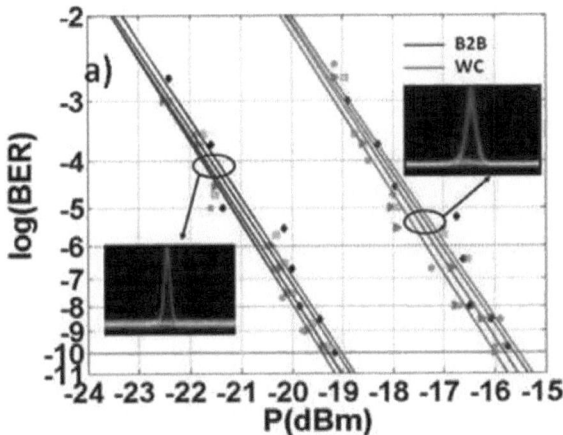

Fig. 7.15 Bit-error rate of back-to-back (B2B) and non-inverted wavelength converter signal (WC) [59].

Fig. 7.15 depicts the BER curves of the back-to-back (B2B) and non-inverted wavelength converter signal (WC). Error-free operation was achieved with power penalties less than 4 dB. The rather high penalty was attributed to the low power and the reduced quality of the signal after the carrier suppression and amplification. A SEM inspection showed a 2 μm lateral misalignment between the SOA facet and SOI waveguide. Measurements of the other recently hybrid integrated AOWCs should be able to improve the wavelength conversion performance by 1-2 dB penalty reduction.

8 Summary

This PhD thesis describes the design and characterization of Mach-Zehnder delay interferometer (MZ-DI) in 4 µm SOI technology. This includes the realization of the following components: straight & bent rib waveguides, multi-mode interference couplers (MMI) and integrated heaters.

In a first step, the dimensions for single-mode straight & bent rib waveguides were determined by doing simulation and performing experiments. After iterations in fabrication, straight waveguides with low loss (~ 0.15 dB/cm) and minimum bending radii of (~ 5 mm) were realized.

In a second step, detailed MMI coupler design considerations in SOI material were carried out. In particular for 2×2 and 4×4 MMI coupler in 4µm SOI material, optimum dimensions were determined with respect to low excess loss, imbalance, polarization dependent loss and phase accuracy. The key parameters for the MMI coupler design are the width of the multi-mode section and the width of the access waveguides. The afterwards realized MMI couplers show state-of-the-art performance compared to other PLC technologies, e.g. in Silica or InP. The realized 4×4 MMI couplers in 4µm SOI material show relatively low excess loss (< 0.6 dB) compared to other technologies. The analysis of tolerances given by device fabrication shows flexibility to compensate arising deviations in device geometry. Except for the case of clearly thicker substrates than the nominal 4 µm, an adaption of the etch depth may be required.

Finally, the MMI couplers were included into different MZ-DI designs. The different designs showed high performances with extinction ratios of ~ 30 dB, insertion loss of < 3dB (including coupling loss plus on-chip loss), low imbalance and PDL. Under use of a stress-inducing cladding (with influence on the rib waveguide birefringence), the PDFS could be reduced in each design to < 1 GHz (or < 8 pm). With Omega-design reduction of up to ± 0.21 GHz (under use of TE/TM polarized light) over C-band could be measured. Integrated heaters allow the thermal control of the MZ-DI with a power of 170 mW for π-shift.

The 2×4 MZ-DI, using 2×2 and 4×4 MMI couplers, show the expected power and phase behavior. Nevertheless, the requirement of very low PDFS (± 0.125 GHz) over C-band for DQPSK signal demodulation could not be satisfied. Here, limitations of SOI technology become visible. The high efforts in silica technology in terms of polarization control result in low PDF-shifts in agreement with DQPSK demodulator specifications.

In principle such low PDFS can also be reached in SOI material. A compensation techniques different from stress induced birefringence tuning is required.

Dynamic measurements (eye diagrams, BER measurements) at the Fraunhofer Institut Berlin (HHI) provided the evidence of accurate MZ-DI operation for DPSK signal demodulation in real data systems. Here, the 4 μm SOI technology is able to match the performances of other DPSK demodulators realized in other technologies. Low insertion losses of fiber based devices can not be reached with PLC technology.

Within the frame of BMBF and European projects (EIBONE, BOOM), the MZ-DIs were also combined with III/V material component, i.e. a balanced photo detector (BPD) and semiconductor optical amplifiers (SOA). Here, the hybrid integration approach of TU-Berlin could be tested successfully. The flip-chip process was further improved in the BOOM project by use of flip-chip capabilities at TU Valencia. However, it requires still additional effort to minimize coupling losses between integrated component and SOI board waveguide.

The DPSK demodulator with integrated balanced photo detector was completed with a package in cooperation with "u2t-photonics". This package passed system tests quite successfully. Packages of the integrated AOWC are ordered and were delivered in September 2011. Packaged passive AOWC devices were tested successfully.

Finally, the SOI technology offers their high potential for realization of high-performance devices. The hybrid integration on SOI-board enables furthermore the creation of compact devices, which combine optics and electronics in an innovative way.

9 List of acronyms

AOWC	all-optical wavelength converter
ARC	anti-reflection coating
BER	bit-error rate
BOX	buried oxide
BPD	balanced photo detector
B2B	back-to-back
C-band	conventional band (1530nm – 1565nm)
CW	continuous wave
DPSK	differential phase-shift keying
DQPSK	differential quadrature phase-shift keying
EDFA	erbium-doped fiber amplifier
EIBONE	Efficient Integrated Backbone (BMBF project)
FDM	finite difference mode
FMLL	fiber mode-locked laser
F2F	fiber-to-fiber
FSR	free spectral range
MM	multi-mode
MMI	multi-mode interference
MFD	mode-field diameter
MZ-DI	Mach-Zehnder delay interferometer
NRZ	non-return-to-zero
OBF	optical bandpass filter
OOK	on-off keying
OSNR	optical signal-to-noise ratio

PC	physical contact
PLC	planar lightwave circuit
PRBS	pseudo random bit sequence
RIE	reactive-ion etch
RF	radio frequency
RZ	Return-to-zero
SEM	scanning electron microscope
SOA	semiconductor optical amplifier
SOI	silicon-on-insulator
SM	single-mode
SOP	state-of-polarization
TE	transverse-electric
TLS	tunable laser source
TM	transverse-magnetic

10 List of symbols

a_i	coupling coefficient of the i-th MMI coupler
b_i	coupling coefficient of the i-th MMI coupler
$C_{1,2}$	stress optic constant (1/Pa)
c	speed of light in vaccum (m/s)
d	half width of guiding layer H (µm)
$\underline{E}_{i1,i2,2x2}$	electrical input fields of the 2×2 MZ-DI (V/m)
$\underline{E}_{o1,o2,2x2}$	electrical output fields of the 2×2 MZ-DI (V/m)
$\underline{E}_{i1,i2,2x4}$	electrical input fields of the 2×4 MZ-DI (V/m)
$\underline{E}_{o1,o2,o3,o4,2x4}$	electrical output field of the 2×4 MZ-DI (V/m)
EL	excess loss (dB)
ED	etch depth (µm)
ER	extinction ratio (dB)
$ER_{o,2x2}$	extinction ratios at 2×2 MZ-DI outputs (dB)
FSR	free-spectral range (nm)
f	frequency (Hz)
G'_{th}	Length-dependent thermal conductance (m/Ω)
H	SOI substrate height (µm)
H_{SiO2}	buried oxid layer thickness (µm)
h	slab height (µm)
IL	insertion loss (dB)
IMB	imbalance (dB)
I_{SOA}	SOA driving current (mA)
k_0	free-space wave number (1/µm)
$k_{x,m}$	lateral wave number of the m-th mode (1/µm)

L	interferometer arm length (μm)
L_H	heater length (μm)
L_{mmi}	MMI-coupler length (μm)
L_{th}	thermal diffusion length (μm)
n_{eff}	effective refractive index
n_{eff}^{TE}	effective refractive index of fundamental TE mode
n_{eff}^{TM}	effective refractive index of fundamental TM mode
n_c	effective refractive index of fundamental mode in a planar waveguide
$\overline{n_g}$	averaged group index
n_g^{TE}	group index TE mode
n_g^{TM}	group index TM mode
n_r	effective refractive index of fundamental mode in the surrounding material (cladding) of a planar waveguide
n_{Si}	refractive index silicon
n_{SiO2}	refractive index silicon dioxide
$n_{x,y}$	refractive index in x, y direction
P	power (W)
$P_{o1,o2,o3,o4,2x2}$	output power of the 2×2 MZ-DI (W)
$P_{o1,o2,o3,o4,2x4}$	output power of the 2×4 MZ-DI (W)
$PDFS$	polarization dependent frequency shift (GHz)
PDL	polarization dependent loss (dB)
$PD\lambda\text{-}shift$	polarization dependent wavelength shift (pm)
P_{laser}	cw-output power laser source (dBm)
P_{ref}	reference signal (dB)
P_{tot}	normalized total guided mode power

P_π	heater power for π phase shift (W)
R	radius (μm)
R'_{th}	length-dependent resistance (Ω/m)
r	ratio h/H
t	cladding thickness (μm)
T	temperature (K)
TEC	thermal expansion coefficient (1/K)
TOC	thermo-optic coefficient (1/K)
$\underline{T}_{\Delta\varphi}$	transfer matrix of phase shift section
$\underline{T}_{MMI,2x2}$	transfer matrix of 2×2 MMI coupler
$\underline{T}_{MMI,4x4}$	transfer matrix of 4×4 MMI coupler
$\underline{T}_{MZ-DI,2x2}$	transfer matrix of 2×2 MZ-DI
$\underline{T}_{MZ-DI,2x4}$	transfer matrix of 2×4 MZ-DI
V_{SOA}	SOA driving voltage (V)
W	rib waveguide width (μm)
$W_{H,eff}$	effective heater width (μm)
W_H	heater width (μm)
W_{inp}	width input waveguide (μm)
W_{mmi}	MMI-coupler width (μm)
$W_{mmi,eff}$	effective MMI coupler width (μm)
α_{th}	thermal decay constant (1/μm)
β_0	propagation constant of the fundamental mode (1/μm)
β_1	propagation constant of the 1st order mode (1/μm)
β_m	propagation constant of the m-th mode (1/μm)
$\beta_{0,height\_h}$	propagation constant of fundamental planar waveguide mode with height h (1/μm)

$\beta_{0,height\_H}$	propagation constant of fundamental planar waveguide mode with height H (1/μm)
β_0^{TE}	propagation constant of fundamental TE mode (1/μm)
β_0^{TM}	propagation constant of fundamental TE mode (1/μm)
γ_{th}	thermal expansion coefficient (1/K)
Δf	polarization dependent frequency shift = PDFS (GHz)
$\Delta \varphi$	phase shift (rad)
$\Delta \varphi_{o1,o2,o3,o4}$	differential phase shift at MMI coupler outputs (rad)
ΔH	substrate thickness variation (μm)
$\Delta (H-h)$	etch depth variation (μm)
ΔL	delay length (μm)
ΔL_H	differential heater length (μm)
$\Delta \lambda_{stress}$	stress-induced wavelength shift (pm)
Δn	change of refractive index by thermo-optic effect
Δn_{eff}	birefringence
Δn_{geo}	geometrical birefringence
$\Delta (\Delta n_{geo})$	change of geometrical birefringence
Δn_{stress}	stress-induced birefringence
ΔT	temperature difference (K)
ΔW	rib waveguide width variation (μm)
$\varphi_{o1,o2,o3,o4}$	phases at MZ-DI output ports (rad)
λ	wavelength (μm)
λ_{FSR}	free-spectral range (pm)
$\lambda_{min.P}$	wavelength at transmission minimum (nm)
λ-shift	temperature dependent wavelength shift (nm)
σ	surface roughness (μm)

$\sigma_{th,Si}$	thermal conductivity silicon (W/m K)
$\sigma_{th,SiO2}$	thermal conductivity silicon oxide (W/m K)
$\sigma_{x,y,z}$	stress tensor components in x,y,z direction (Pa)

11 List of Publications

[1] M. Schnarrenberger, L. Zimmermann, T. Mitze, K. Voigt, J. Bruns, K. Petermann, Concept for an Alternative Solder-Free Flip-Chip Technique on SOI using Black-Silicon, *3rd IEEE International Conference on Group IV Photonics, ThB3*, 2006, pp. 191-193.

[2] M. Schnarrenberger, L. Zimmermann, T. Mitze, K. Voigt, J. Bruns, K. Petermann, Low Loss Star Coupler Concept for AWGs in Rib Waveguide Technology, *IEEE Photonics Technology Letters (PTL), vol.18, iss.23*, December 2006, pp.2469-2471.

[3] K. Voigt, L. Zimmermann, G. Winzer, M. Schnarrenberger, T. Mitze, J. Bruns, K. Petermann, Silicon-on-Insulator (SOI) Delay-Line Interferometer with Low Polarization-Dependent Wavelength Shift, *European Conference on Integrated Optics (ECIO) Copenhagen, ThC7*, April 25-27, 2007.

[4] J. Bruns, T. Mitze, L. Zimmermann, K. Voigt, M. Schnarrenberger, K. Petermann, An Optical Board Approach Based on SOI (silicon-on-insulator*)*, *9th International Conference on Transparent Optical Networks (ICTON) 2007, WE.A1.1*, July 2007, pp. 179-182.

[5]. K. Voigt , L. Zimmermann , G. Winzer , T. Mitze , J. Bruns , K. Petermann, C. Schubert SOI Delay Interferometer with Tuned Polarization Dependent Wavelength Shift for 40 Gbit/s DPSK Demodulation, *4th IEEE International Conference on Group IV Photonics, WA4*, 2007, pp. 10-12.

[6]. L. Zimmermann, K. Voigt, G. Winzer, J. Bruns, and K. Petermann, Optimization Considerations for 4 µm SOI-Waveguide Technology with Respect to Polarization Dependence, *4th IEEE International Conference on Group IV Photonics, WP43*, 2007, pp. 174-176.

[7]. L. Zimmermann, K. Voigt, G. Winzer, T. Mitze, J. Bruns, K. Petermann, T. Richter, C. Schubert, Silicon-on-Insulator (SOI) Delay-Line Interferometer with Low Polarization-Dependent Frequency Shift for 40 Gbit/s DPSK Demodulation, *33rd European Conference on Optical Communication (ECOC) Berlin, 7.3.4*, Sept. 2007.

[8]. K. Voigt, L. Zimmermann, G. Winzer, T. Mitze, J. Bruns, K. Petermann, B. Hüttl, C. Schubert, Performance of 40 Gbit/s DPSK Demodulator in SOI-Technology, *IEEE Photonics Technology Letters (PTL), vol.20, iss.8*, April 2008, pp. 614-616.

[9]. K. Voigt, L. Zimmermann, G. Winzer, K. Petermann, SOI based 2×2 and 4×4 waveguide couplers - Evolution from DPSK to DQPSK, *5th IEEE International Conference on Group IV Photonics, ThB5*, 2008, pp. 209-211.

[10]. L. Zimmermann, K. Voigt, D. Wolansky, H. Richter, B. Tillack, Silicon photonics front-end integration in high-speed 0.25μm SiGe BiCMOS, *5th IEEE International Conference on Group IV Photonics, FB5*, 2008, pp. 374-376.

[11]. K. Voigt, L. Zimmermann, G. Winzer, K. Petermann, C.M. Weinert, Silicon-on-insulator 90°optical hybrid using 4x4 waveguide couplers with C-band operation, *34th European Conference on Optical Communication (ECOC) Brussels, Tu.3.C.5*, Sept. 2008.

[12]. G. Unterbörsch, M. Kroh, J. Honecker, A.G. Steffan, G. Tsianos, H.-G. Bach, J. Kreissl, R. Kunkel, G.G. Mekonnen, W. Rehbein, D. Schmidt, J. Bruns, T. Mitze, K. Voigt, L. Zimmermann, Hybrid Flip-Chip Integration of a 40 Gb/s DPSK Receiver Comprising a Balanced Photodetector on a DLI-SOI Board, *34th European Conference on Optical Communication (ECOC) Brussels, P.2.O4*, Sept. 2008.

[13]. K. Voigt, L. Zimmermann, J. Bruns, J. Kreissl, M. Kroh, T. Mitze, K. Petermann, 40 Gbit/s DPSK receiver on silicon-on-insulator (SOI), *Workshop on TUNABLE AND ACTIVE SILICON PHOTONICS*, Hamburg, 28.09.-30.09.2008.

[14]. L. Zimmermann, K. Voigt, G. Winzer, K. Petermann, C.M. Weinert, C-band optical 90°-hybrids based on silicon-on-insulator 4×4 waveguide couplers, *IEEE Photonics Technology Letters (PTL), vol.21, iss.3*, Feb. 2009, pp. 143 – 145.

[15]. M. Kroh, G. Unterbörsch, G. Tsianos, R. Ziegler, A.G. Steffan, H.G. Bach, J. Kreissl, R. Kunkel, G.G. Mekonnen, W. Rehbein, D. Schmidt, R. Ludwig, K. Petermann, J. Bruns, T. Mitze, K. Voigt, L. Zimmermann, Hybrid Integrated 40Gb/s DPSK Receiver on SOI, *Optical fiber communication conference (OFC)*, San Diego, OMK3, March 2009.

[16]. K. Voigt, L. Zimmermann, G. Winzer, T. Mitze, K. Petermann, J. Kreissl, E. Tangdiongga, K. Vyrsokinos, L.Stampoulidis, SOI platform for high speed all optical wavelength conversion, *6th IEEE International Conference on Group IV Photonics, ThC5*, 2009, pp. 101-103.

[17]. F. Gomez-Agis, O. Raz, E.J. Zhang, E. Tangdiongga, L. Zimmermann, K. Voigt, C. Vyrsokinos, L. Stampoulidis, H.J.S.Dorren, All-optical wavelength conversion at 160 Gbit/s using an SOA and a silicon-on-insulator photonic circuit, *Electronics Letters, vol. 45, iss.22*, 2009, pp. 1132-1132.

[18]. L. Zimmermann, K. Voigt, G. Winzer, K. Petermann, Towards Silicon on Insulator DQPSK Demodulators, *Invited Talk, Optical Fiber Communication Conference (OFC), OThB3*, San Diego (California), March 2010.

[19]. K. Vyrsokinos, L. Stampoulidis, F. Gomez-Agis, K. Voigt, L. Zimmermann, T. Wahlbrink, Z. Sheng, D.V. Thourhout, H.J.S. Dorren, Ultra-high Speed, All-Optical Wavelength Converters Using Single SOA and SOI Photonic Integrated Circuits, *Photonics Society Winter Topicals Meeting Series (WTM), WC2.3*, 2010, pp. 113-114.

[20]. L. Stampoulidis, K. Vyrsokinos, K. Voigt, L. Zimmermann, F. Gomez-Agis, H. Dorren, Z. Sheng, D. van Thourhout, L. Moerl, J. Kreissl, B. Sedighi, A. Pagano and E. Riccardi, The European BOOM project: Silicon photonics for high-capacity optical packet routers, *IEEE Journal of selected Topics in Quantum Electronics, vol. PP, iss.99*, 2010, pp.1-12.

[21]. K. Voigt, L. Zimmermann, G. Winzer, K. Petermann, New Design Approach to MMI-Couplers in Photonic Wire Substrates, *15th European Conference on Integrated Optics (ECIO), WeP 20*, Cambridge, April 2010.

[22] M. Kroh, L. Zimmermann, H.-G. Bach, A. Beling, M. L. Nielsen, K. Voigt, R. Ludwig, J. Bruns, G. Unterbörsch, Integrated Receivers on SOI for Advanced Modulation Formats, *The Institution of Engineering and Technology (IET), vol.5, iss.5*, 2011, pp.211-217.

[23] L. Zimmermann, K. Voigt, G. Winzer, K. Landles, J. Lynn, S. Duffy, Packaging of SOI motherboards for high-speed all optical router applications, *7th IEEE International Conference on Group IV Photonics, P2.20*, Beijing, Sept. 2010, pp. 281-283.

[24] C. Stamatiadis, F. Gomez-Agis, I. Lazarou, L. Stampoulidis, H.J.S Dorren, L. Zimmermann, K. Voigt, D. Van Thourhout, P. De Heyn, H.Avramopoulos, The BOOM project: Towards 160Gb/s Packet Switching Using SOI Photonic Integrated Circuits and Hybrid Integrated Optical Flip-Flops, *Journal of Lightwave Technology (JLT), vol.30., no.1,* Jan 2011, pp. 22-20.

[25] K. Voigt, L. Zimmermann, G. Winzer, H. Tian, B. Tillack, K. Petermann, Fully passive Si-photonic 90° hybrid for coherent receiver applications, *European Conference and Exposition on Optical Communications (ECOC)*, Geneva, 18-22 Sept. 2011, Switzerland, Tu.3.LeSaleve.3., 2011, pp. 1-3.

[26] K. Voigt, L. Zimmermann, G. Winzer, H. Tian, B. Tillack, K. Petermann, C-band Optical 90° Hybrid in Silicon Nanowaveguide Technology, *IEEE Photonics Technology Letters (PTL), vol.23, iss.23,* 2011, pp.1769-1771.

[27] L. Zimmermann, G.B. Preve, K. Voigt, G. Winzer, J. Kreissl, L. Moerl, C. Stamatiadis, High-precision flip-chip technology for alloptical wavelength conversion using SOI photonic circuit, *8th IEEE International Conference on Group IV Photonics (GFP)*, 14-16 Sept. 2011, London, P2.4, pp. 237-239.

[28] C. Stamatiadis, L. Stampoulidis, K. Vyrsokinos, I. Lazarou, L. Zimmermann, K. Voigt, L. Moerl, J. Kreissl, B. Sedighi, Z. Sheng, P. De Heyn, D. van Thourhout,M. Karl, T. Wahlbrink, J. Bolten, A. Leinse, R. Heidemann, F. Gomez-Agis, H.J.S. Dorren, , A. Pagano, E. Riccardi, H. Avramopoulos, The ICT-BOOM project: Photonic routing on a silicon-on-insulator hybrid platform, *15th International Conference on Optical Network Design and Modeling, ONDM 2011,* 8.-10. Feb. 2011, Bologna, 2011

[29] I. Lazarou, C. Stamatiadis, B. Schrenk, L. Stampoulidis, L. Zimmermann, K. Voigt, G.B. Preve, L. Moerl, J. Kreissl and H. Avramopoulos, Colorless ONU with Discolored Source and Hybrid SOI Integrated Wavelength Converter, *IEEE Photonics Technology Letters, vol.24, iss.5,* March, 2012.

[30] I. Lazarou, C. Stamatiadis, B. Schrenk, L. Stampoulidis, L. Zimmermann, K. Voigt, G. Preve, L. Moerl, J. Kreissl, H. Avramopoulos, Migrating Legacy PON Equipment towards Colorless ONU through Hybrid Integrated SOI All-Optical Wl-Converter, *Optical Fiber Communication Conference (OFC), OM2I.1,* March 4-8, 2012, Los Angeles, California.

[31] A. Rahim, S. Schwarz, J. Bruns, K. Voigt, D. Kroushkov, M. Arnous, L. Zimmermann, C. Schaeffer, K. Petermann, Tunable Residual Dispersion Compensator Using Generalized MZIs In Silicon-on-Insulator, *Optical Fiber Communication Conference (OFC), OTh4D.4,* March 4-8, 2012, Los Angeles, California.

[32] C. Stamatiadis, L. Stampoulidis, K. Vyrsokinos, I. Lazarou, L. Zimmermann, K. Voigt, G. Preve, L. Moerl, L. Moerl, J. Kreissl, H. Avramopoulos, A Hybrid Photonic Integrated Wavelength Converter on a Silicon—on-Insulator Substrate, *Optical Fiber Communication Conference (OFC), OM3E,* March 4-8, 2012, Los Angeles, California.

[33] A. Rahim, S. Schwarz, J. Bruns, K. Voigt, Kroushkov, D., Arnous, M.H., Schaffer, C.G., K. Petermann, Finite Impulse Response Filter using 4-port MMI couplers for Residual Dispersion Compensation, *IEEE Journal of Lightwave Technology (JLT), vol.pp, iss.99,* Jan. 2012.

12 Bibliography

[1]. **Baets, R. et al.** Silicon-on-Insulator based Nano-photonics: Why, How, What for? *International Conference on Group for Photonics (GFP), Plenary session FA1.* 2005, pp.168-170.

[2]. **Mitze, T., Schnarrenberger, M., Zimmermann, L., Bruns, J., Fidorra, F., Janiak, K., Kreissl, J., Fidorra, S., Heidrich, H., Petermann, K.** CWDM Transmitter Module Based on Hybrid Integration. *IEEE Journal of Selected Topics in Quantum Electronics, Vol.12, No.5.* Oct. 2006, pp.983-987.

[3]. **Mitze, T., Schnarrenberger, M., Zimmermann, L., Bruns, J., Fidorra, F., Kreißl, J., Janiak, K., Fidorra, S., Heidrich, H., Petermann, K.** Hybrid Integration of II/V Lasers on a Silicon-on-Insulator (SOI) Optical Board. *1st IEEE International Conference on Group IV Photonics, WA1 (Plenary).* Hong Kong, China, 2004.

[4]. **Bestwick, T.** ASOCtm - a Silicon-Based Integrated Optical Manufacturing Technology. *48th Electronic Components and Technology Conference, pp.566-571.* Seattle WA, USA, 1998.

[5]. **House, A., Whiteman, R., Kling, L., Day, S., Knigths, A., Hogan, D., Hopper, F., Asghari, M.** Silicon Waveguide Integrated Optical Switching with Microsecond Switching Speed. *OFC Atlanta / Georgia 2003, ThD5.* 2003.

[6]. **Day, I., Evans, A., Knigths, A., Hopper, F., Roberts, S., Johnston, J., Day, S., Luff, J., Tsang, H., Asghari, M.** Tapered Silicon Waveguides for Low Indertion Loss Highly-Efficient High-Speed Electronic Variable Optical Attenuators. *OFC Atlanta / Georgia 2003, TuM5.* 2003.

[7]. **Fischer, U., Zinke, T., Kropp, J.-R., Arndt, F., Petermann, K.** 0.1 dB/cm Waveguide Losses in Single-Mode SOI Rib Waveguides. *IEEE Photonics Technology Letters, Vol.8, No.5.* May 1996, pp. 647-648.

[8]. **Fischer, U., Zinke, T., Schüppert, B., Petermann, K.** Singlemode optical switches based on SOI waveguides with large cross section. *Electronics Letters, Vol.30, No.5.* March 1994, pp.406-408. .

[9]. **Winzer, P.J., Essiambre, R.-J.** "Advanced Optical Modulation Formats". *Proceedings of the IEEE, Vol.94, No.5.* May 2006, pp. 952-985.

[10]. **Gnauck, A.H., Winzer, P.J.** Optical Phase-Shift-Keyed Transmission. *IEEE Journal of Lightwave Technology, Vol.23, No.1.* Janaury 2005, pp. 115-130.

[11]. **Madsen, C.K., Zhao, J.H.** *Optical Filter Design and Analysis - A signal Processing Approach.* s.l. : Wiley & Sons, 1999.

[12]. **Ducournau, G., Latry, O., Ketata, M.** The All-fiber MZI Structure for Optical DPSK Demodulation and Optical PSBT Encoding. *Journal of Systemics, Cybernetics and Informatics, Vol.4, No. 4.* 2006, pp.78-89.

[13]. **Bachmann, M., Besse, P.A., Melchior, H.** General self-imaging properties in NxN multimode interference couplers including phase relations. *Applied Optics, Vol.33, No.18.* June 1994, pp. 3905-3911.

[14]. **Kim, H. und Winzer, P.J.** Robustness to Laser Frequency Offset in Direct-Detection DPSK and DQPSK Systems. *IEEE Journal of Lightwave Technology, Vol.21, No.9.* September 2003, pp. 1887-1891.

[15]. **Winzer, P.J., Kim, H.** Degradations in Balanced DPSK Receivers. *IEEE Photonics Technology Letters, Vol.15, No.9.* September 2003, pp. 1282-1284.

[16]. **Bosco, G., Poggiolini, P.** The Effect of Receiver Imperfections on the Performance of Direct-Detection Optical Systems using DPSK Modulation. *Proceedings OFC 2003, ThE6.* 2003, pp. 457-458.

[17]. **Liu, X., Gnauck, A.H., Wei, X., Hsieh, J.Y.C., Ai, C., Chien, V.** Athermal Optical Demodulator for OC-768 DPSK and RZ-DPSK Signals. *IEEE Photonics Technology Letters, Vol.17, No.12.* December 2005, pp. 2610-2612.

[18]. **Lize, Y.K., Faucher, M., Jarry, E., Oulette, P., Villeneuve, E., Wetter, A., Seguin, F.** Phase-Tunable Low-Loss, S-, C-, and L-Band DPSK and DQPSK Demodulator. *IEEE Photonics Technology Letters, Vol.19, No.23.* December 1, 2007, pp. 1886-1888.

[19]. **Doerr C.R., Gill, D.M., Gnauck, A.H., Buhl, L.L., Winzer, P.J., Capuzzo, M.A., Wong-Foy, A., Chen, E.Y., Gomez, L.T.** Monolithic Demodulator for 40-Gb/s DQPSK Using a Star Coupler. *IEEE Journal of Lightwave Technology, Vol. 24, No.1.* 1. January 2006, pp.171-174.

[20]. **Oguma, M., Nasu, Y., Takahashi, H., Kawakami, H., Yoshida, E.** Single MZI-based 1x4 DQPSK demodulator. *European Conference on Optical Communication (ECOC) 2007, Berlin, Germany, Thursday 10.3.3.* 16-20 September 2007.

[21]. **Soref, R.A., Schmidtchen, J., Petermann, K.** Large Single-Mode Rib Waveguides in GeSi-Si and Si-on-SiO2. *IEEE Journal of Quantum Electronics, Vol.27, No.8.* August 1991, pp. 1971-1974.

[22]. **Petermann, K.** Properties of Optical Rib-Guides with Large Cross-Section. *Archiv für Elektronik und Übertragungstechnik (AEÜ), Band 30, Heft 3.* 1976, pp. 139-140.

[23]. **Aalto, T., Harjanne, M.** New Single-Mode Condition for Silicon Rib Waveguides . *12th European Conference on Integrated Optics (ECIO), Grenoble.* 6-8 April 2005.

[24]. **Rickman, A.G., Reed, G.T., Namavar, F.** Silicon-on-Insulator Optical Rib Waveguide Loss and Mode Characteristics. *Journal of Lightwave Technology, Vol.12, No.10.* October 1994, pp. 1771-1776.

[25]. **Zinke, T., Fischer, U., Splett, A., Schüppert, B., Petermann, K.** Comparison of optical waveguide losses in silicon-on-insulator. *Electronics Letters, Vol.29, No.23.* November 1993, pp. 2031-2033.

[26]. **Payne, F.P., Lacey, J.P.R.** A theoretical analysis of scattering loss from planar optical loss. *Optical and Quantum Electronics 26.* 1994, pp. 977-986.

[27]. **Marcatili, E.A.J., Miller, S.E.** Improved relationships describing directional control in electromagnetic wave guidance. *Bell Syst. Tech.J., 48.* 1969, pp.2161-2188.

[28]. **Kitoh, T., Takato, N., Yasu, M., Kawachi, M.** Bending Loss Reduction in Silica-Based Waveguides by Using Lateral Offsets. *Journal of Lightwave Technology, Vol.13, No.4.* April 1995, pp.555-562.

[29]. **Schnarrenberger, M.** *Optische Filter aus Silizium-Rippenwellenleitern.* Dissertation, Fak. IV (Elektrotechnik & Informatik) D83, TU-Berlin, Sept. 2008.

[30]. **Dai, D., He, S.** Analysis of the birefringence of a silicon-on-insulator rib waveguide. *Applied Optics, Vol.43, No. 5.* February 2004, pp. 1156-1161.

[31]. **Huang, M.** Stress effects on the performance of optical waveguides. *International Journal of Solids and Structures 40.* 2003, pp. 1615-1632.

[32]. **Xu, D.-X., Cheben, P., Dalacu, A., Delage, A., Janz, S., Lamontagne, B., Picard, M.-J., Ye, W.N.** Eliminating the birefringence in silicon-on-insulator ridge waveguides by use of cladding stress. *Optics Letters, Vol.29, No.20.* October 2004, pp. 2384-2386.

[33]. **Xu, D.-X., Cheben, P., Janz, S., Dalacu, D.** Control of SOI waveguide polarization properties for microphotonic applications. *Proc. of CLEO/Pacific RIM 2003, CD-ROM (IEEE, Piscataway, NJ, 2003).* 2003.

[34]. **Nasu, Y., Oguma, M., Takahashi, H., Inoue, Y., Kawakami, H., Yoshida, E.** Polarization insensitive MZI-based DQPSK demodulator with asymmetric half-wave plate configuration. *OFC/NFOEC, OthE5.* 2008.

[35]. **Sakamaki, Y., Nasu, Y., Hashimoto, K. H., Inoue, Y., Takahashi, H.** Silica Waveguide DQPSK Demodulator with Wide Operation Range Enhanced by Using Stress Release Grooves. *IEEE Photonics Technology Letters, Vol.21, No.13.* July 2009, pp.938-940.

[36]. **Bryngdhal, O.** Image formation using self-imaging techniques. *Journal of the Optical Society of America, Vol.63, No.4.* April 1973, pp. 416-419.

[37]. **Ulrich, R.** Image Formation by Phase Coincidences in Optical Waveguides. *Optics Communications, Vol.13, No.3.* March 1975, pp. 259-264.

[38]. **Soldano, L.B., Pennings, E.C.M.** Optical Multi-Mode Interference Devices Based on Self-Imaging: Principles and Applications. *IEEE Journal of Lightwave Technology, Vol.13, No.4.* April 1995, pp. 615-627.

[39]. **Petermann, K.** *Vorlesungsskript Hochfrequenztechnik I.* 2004.

[40]. **Ulrich, R., Kamiya, T.** Resolution of self-images in planar optical waveguides. *Journal Opt. Society, Vol.68, No.5.* May 1978, 583-592.

[41]. **Kaalund, C.J., Jin, Z.** Novel multimode interference devices for low index contrast material systems featuring deeply etched air trenches. *Optics Communications 250.* 2005, pp. 292-296.

[42]. **Shi, Y., Dai, D.** Design of a compact multimode interference coupler based on deeply-etched SiO2 ridge waveguides. *Optics Communications 271.* 2007, pp. 404-407.

[43]. **Besse, P.A., Bachmann, M., Melchior, H.** Phase relations in multi-mode interference couplers and their application to generalized integrated Mach-Zehnder optical switches. *European Conference on Integrated Optics (ECIO), Neuchatel, Switzerland.* April 1993, pp. 2/22-2/23.

[44]. **Besse, P. A., Bachmann, M., Melchior, H., Soldano, L.B., Smit, M.K.** Optical Bandwidth and Fabrication Tolerances of Multimode Interference Couplers. *Journal of Lightwave Technology, Vol.12, No.6.* June 1994, pp. 1004-1009.

[45]. **Wei, H., Yu, J., Liu, Z., Zhang, X., Shi, W., Fang, C.** Fabrication of 4x4 Tapered MMI Coupler with Large Cross Section. *IEEE Photonics Technology Letters, Vol.13, No.5.* May 2001, pp. 466-468.

[46]. Pennings, E.C.M., Deri, R.J., Bhat, R., Hayes, T.R., Andreadakis, N.C. Ultracompact, All-Passive Optical 90°-Hybrid on InP Using Self-Imaging. *IEEE Photonics Technology Letters, Vol.5, No.6.* June 1993, pp. 701-703.

[47]. **Heidrich, H., Hoffmann, D., Döldissen, W., Klug, M.** Integrated Optical 4x4 Star Coupler on LiNbO3. *Electronics Letters, Vol.20, No. 25/26.* Dec. 1984, pp. 1058-1059.

[48]. **Li, H.H.** Refractive Index of Silicon and Germanium and Its Wavelength and Temperature Derivatives . *J. Phys. Chem. Ref. Data, Vol.9, No.3.* 1980, pp.561-658.

[49]. **El-Bawab, T.S.** *Optical Switching.* New York : Springer Science+Business Media, Inc., 2006. 0-387-26141-9.

[50]. **Fischer, U.** Entwicklung und Optimierung eines integriert-optischen Schaltmoduls in Silizium. *Dissertation D83, TU-Berlin.* 1995.

[51]. **Beling, A.** PIN Photodiodes Modules for 80Gb/s and Beyond. *Optical Fiber Communication Conference (OFC), Anaheim, California, OFI1.* March 5, 2006.

[52]. **Kroh., M., Unterbörsch, G., Tsianos, G., Ziegler, R., Steffan, A.G., Bach, H.-G., Kreissl, J., Kunkel, R., Mekonnen, G.G., Rehbein, W., Schmidt, D., Ludwig, R., Petermann, K., Bruns, J., Mitze. T., Voigt, K., Zimmermann, L.** Hybrid Integrated 40Gb/s DPSK Receiver on SOI. *Optical fiber communication conference (OFC), San Diego, OMK3.* March 2009.

[53]. **Liu, Y., Tangdiongga, E., Li., Z., de Waardt, H., Koonen, A.M.J., Khoe, G.D., Dorren, H.J.S., Shu, X., Bennion, I.** Error-free 320 Gb/s SOA-based Wavelength Conversion using Optical Filtering. *Journal of Lightwave Technology (JLT), Vol.25, Iss.1.* 2007, pp.103-108.

[54]. **Nakamura, S., Ueno, Y., Tajima, K.** 168 Gb/s all-optical wavelength conversion with a symmetric-Mach-Zehnder-Type switch. *IEEE Photonics Technology Letters, Vol. 13, Iss.10.* 2001, pp.1091-1093.

[55]. **Leuthold, J., Moller, L., Jaques, J., Cabot, S., Zhang, L., Bernascani, P., Capuzzo, M., Gomez, L., Laskowski, E., Chen, E., Wong-Foy, A., Griffin, A.** 160 Gb/s SOA all-optical wavelength converter and assessment of its regenerative properties . *Electronics Letters (EL), Vol.40, Iss.9.* 2004, pp.554-555.

[56]. **Bernascani, P., Zhang, L., Yang, W., Sauer, L., Buhl, L.L., Sinsky, J.H., Kang, I., Chandrasekhar, S., Neilson, D.T.** Monolithically integrated 40-Gb/s switchable wavelength converter. *Journal of Lightwave Technology (JLT), Vol.24, Iss.1 .* 2006, pp.71-75.

[57]. **Liu, Y., Tangdiongga, E., Zhang, S., de Waardt, H., Khoe, G.D., Dorren, H.J.S.** Error-Free All-Optical Wavelength Conversion at 160 Gb/s Using a Semiconductor Optical Amplifier and an Optical Bandpass Filter. *Journal of Lightwave Technology (JLT), Vol.24, No.1.* Jan. 2006, pp.230-236.

[58]. **Gomez-Agis, F., Raz, O., Zhang, S.J., Tangdiongga, E., Zimmermann, L., Voigt, K., Vyrsokinos, C., Stampoulidis, L., Dorren, H.J.S.** All-optical wavelength conversion at 160Gbit/s using SOA and silicon-on-insulator photonic circuit. *Electronic Letters, Vol.45, No.22.* 22nd October 2009.

[59]. **Stamatiadis, C., Stampoulidis, L., Vyrsokinos, K., Lazarou, I., Kalavrouziotis, D., Zimmermann, L., Voigt, K., Preve, G.B., Moerl, L., Kreissl, J., Avramopoulos.** A Hybrid Photonic Integrated Wavelength Converter on a Silicon-on-Insulator Substrate. *Optical fiber communication conference (OFC), Los Angeles, OM3E.* March 2012.

[60]. **Liu, Y., Tangdiongga, Z., Li, Z., Zhang, S., H. de Waardt, Khoe, G.D., Dorren, H.J.S.** Error-Free All-Optical Wavelength Conversion at 160Gb/s Using a Semiconductor Optical Amplifier and an Optical Bandpass Filter. *Journal of Lightwave Technology, Vol. 24, No.1.* January 2006, pp. 230-236.

13 Acknowledgements

I would like to thank my supervisor Prof. K. Petermann for his guidance during the development of this thesis. I'm grateful Dr. L. Zimmermann for his day-to-day advice.

I would like to thank my colleagues, who supported me at the Institute for High Frequency and Semiconductor System Technologies at the Technische Universität Berlin. Special thanks to E. Brose and G. Winzer.

Thanks to my family, my wife Severine and our children Siri and Juri for their patience and encouraging moments.

i want morebooks!

Buy your books fast and straightforward online - at one of world's fastest growing online book stores! Environmentally sound due to Print-on-Demand technologies.

Buy your books online at
www.get-morebooks.com

Kaufen Sie Ihre Bücher schnell und unkompliziert online – auf einer der am schnellsten wachsenden Buchhandelsplattformen weltweit! Dank Print-On-Demand umwelt- und ressourcenschonend produziert.

Bücher schneller online kaufen
www.morebooks.de

VDM Verlagsservicegesellschaft mbH
Heinrich-Böcking-Str. 6-8
D - 66121 Saarbrücken

Telefon: +49 681 3720 174
Telefax: +49 681 3720 1749

info@vdm-vsg.de
www.vdm-vsg.de

Printed by Books on Demand GmbH, Norderstedt / Germany